From Mosquitoes to Elephants

A Life with Animals

From Mosquitoes to Elephants

A Life with Animals

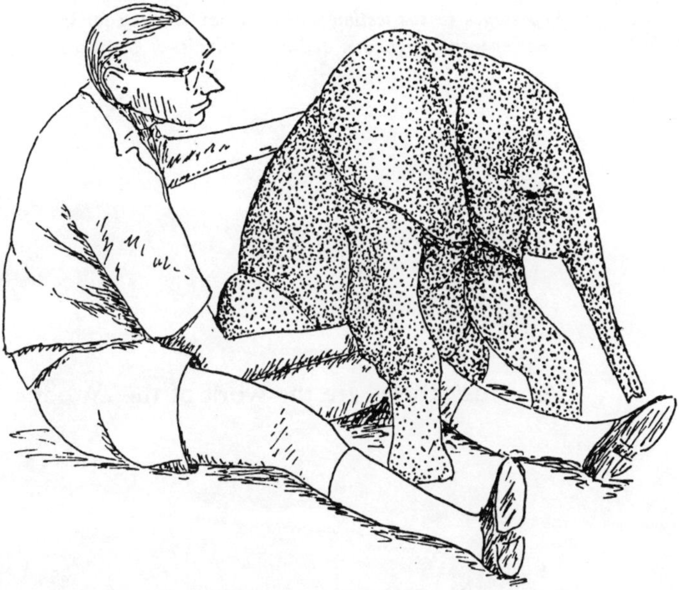

Niels Bolwig

First edition published by Skarv-Nature Publications,
Holte, Denmark © 1983
This edition published by Niels Bolwig,
Ford Cottage, Kilham Lane, Shipton Oliffe,
Cheltenham, Gloucestershire
IN ASSOCIATION WITH
The Self Publishing Association Limited
© 1988
ISBN 1 85421 038 6

All illustrations are the work of the author

Typeset by Printit-Now, Upton upon Severn,
Printed and bound in Great Britain by
Billing & Sons Ltd, Worcester

CONTENTS

Niels Bolwig

Niels Bolwig (b. 1911) graduated from Copenhagen University and was in 1946 awarded his doctor's degree for his work on sense organs in insects. In 1946 he and his family emigrated to South Africa where he had been appointed Senior Lecturer in Zoology. In 1959 he moved to Makerere in Kampala, Uganda. From 1962 he was Professor and Head of Department of Zoology at Ibadan University, Nigeria. Here he organized and expanded a small animal collection belonging to the University into a registered public zoo. In 1966 he left Nigeria for U.S.A. to become the Scientific Curator of Oklahoma City Zoo and a Visiting Professor of Psychology at Oklahoma University. In 1970 he returned to Ibadan University and was retired from there in 1972. Thereafter he ran a small private primate laboratory in Denmark and taught at the University of Copenhagen. In 1977 he was appointed DANIDA-professor of Biology at Roma University, Lesotho. He finally retired in 1979 and is now living in England. Bolwig is the author of a great number of scientific and popular publications spanning a wide field of interest. One book, "My Friend the Vole", has been translated into English.

EARLY DAYS

My lifelong interest in nature began when I was nearly five years old, in a garden at Fortunen, north of Copenhagen, where my parents had rented some rooms for the summer. It must have been early in June, because I remember clearly that the tulips were still in bloom. My grandmother, who was visiting, guided me from flower to flower, showing me their parts – their stamens and their ovaries. She cut through the ovary of a tulip to show me the tiny seeds, which could not grow and develop unless pollen from another tulip landed on the stigma.

My grandmother showed me the bumble bees, furry like my teddy bear, and the smaller honey bees. Every time they entered a flower to lick a small drop of nectar from the base of the colourful petals, they got some pollen on to their fur.

When, a little later, they visited another flower, they brushed against the stigma, leaving some pollen behind them.

Although so young, I listened, fascinated, to all she had to tell me. She told me about the wisdom of the Creator, how wonderfully He had made everything stirring around me. For the first time, the beauty and richness of nature began to dawn on me. New worlds were opening up, and I wanted to know more.

The following year I sat on the door-step and let the mosquitoes bite me. I wanted to find out how they did it. The memory of a mosquito drinking so much that it burst has ever since worried me. Was it my imagination playing a trick on me, or did I really see it? Was anything like that possible?

My interest in nature grew and with that, my love for animals. I found it easy to imagine that animals had characteristics similar to humans, but that plants did not. The following year, we rented rooms where our landlord had an Irish terrier; my brother and I were delighted. We played with the dog, and gave it rides in our wheelbarrow.

The same autumn, my mother's much beloved seamstress, Mrs Nielsen, died. It was the first time that someone close to me had died. I was grief stricken. At the funeral, I had all kinds of thoughts about her in her coffin. Could she hear the music and the hymns? Was she now an angel with wings? I thought I could feel her sitting next to me in the small chapel, telling me not to be sad. Yet I could not help weeping and, as a consolation, I was allowed to inherit her canary. Of course I had no real understanding of animals and their needs, and I knew nothing about how to handle them. The bird now served to satisfy the child's boundless need to love, the same way that the terrier had done during the summer. It was a miracle that my love did not kill it. When there was no one about I took it out of its cage and caressed it.

Next year the canary was given a mate, and we began to breed from them. Later on, different species were added to our collection. The birds were usually allowed to fly freely in our dining room for some hours every day. It must be stressed that my parents were very indulgent; they probably shared my interests to some degree. The same could not always be said about the few visitors who now and again called on my mother for a cup of afternoon coffee. They usually resented it when a bird decided to roost in their hair and spoil their hair-dos.

When I was six years old, my father bought a plot close to where we had stayed the previous summers. Here, with the help of an acquaintance, he built a small wooden cabin. It was small and modest, but it served its purpose as our summer home for many years. To us children it was paradise, much better than our third-floor flat in town.

My father could not have chosen a more suitable place for a budding biologist. There were fields and woods and wonderful ponds. I could easily cover the distance to Klampenborg on the shores of Øresund on my bicycle. There were adventures everywhere. Out in the sandpit were lizards. When I was lying quietly in the grass they would come right up to me. Along the hedgerows between the fields I could find nests: thrushes, black-caps, warblers, nightingales, chaffinches and many other birds.

Most exciting of all, however, was the life in the ponds. I would lie on my stomach and gaze into the brown water. Nothing here prevented me from observing the bloody battles between animals in their struggle for existence. The dark, greenish brown crayfish hid, well camouflaged among the tree roots. Whenever an earthworm was dropped to it, it would grab it and devour the worm bit by bit.

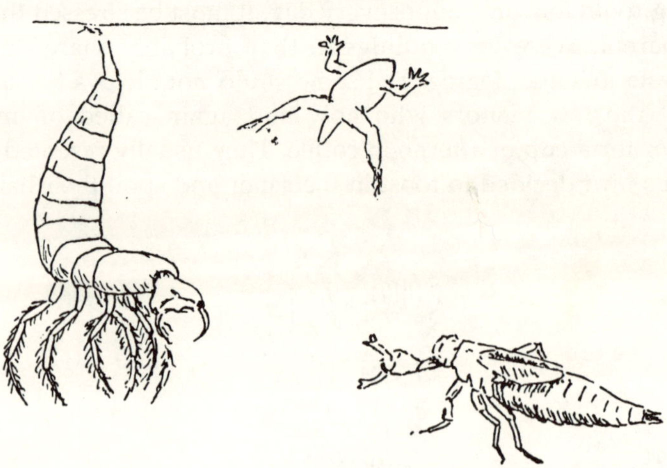

There were also dragonfly larvae with their peculiar "lower lip" which could unfold and be used for grabbing worms and other small prey. Most fearsome of all creatures were the larvae of the water beetles. Their huge, curved, sharp-pointed jaws were penetrated by a canal leading from the closed mouth to the tip of the jaw. Through this canal they could inject into their prey a liquid that dissolved its soft parts. What was dissolved was then sucked up through the same canal and swallowed.

Once I gave to a friend a jam-jar with a large larva of a water beetle. I had also put some tadpoles in the same jar. They had got limbs and were ready to go up on land to live as frogs. Next morning, my horrified friend came and told me the gruesome story; all the tadpoles had died. Their empty bloated skins were floating in the water, looking like "corpses of children". The larva had, during the night, feasted on the tadpoles.

The larvae of the caddis flies were particularly interesting. There were many kinds, each carrying its particular type of dwelling with it wherever it went, as a snail carries its shell. Some larvae made their houses from sand held together by strong threads of silk. Others used straw, sticks or leaves as building materials. I always admired the tidiness with which the houses were built.

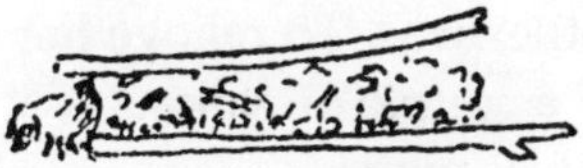

Although jam-jars remained an important container for my freshwater insects, I found them useless when I wanted to keep fish. Then, one Christmas my uncle gave me a small aquarium. With this, new possibilities opened up. Not only could I keep fish and learn about their habits, but I could also, with the help of stones and plants, create settings similar to those I knew from the pond. This fact also taught me something about perspective. Moreover, from then on I was able to have something at home which reminded me of nature in the wild all year round. For hours, I could sit and watch the movements of my fish, and when a guppy gave birth to young ones, I would keep guard until the small hours of the night to prevent them being eaten by their mother.

Gradually, my explorations made me conscious of the countless variations in animals appearance. My father, who was a bookseller, helped me with a variety of books on nature, which I read from cover to cover. I used to get up early before anybody else to read in peace. At night, I often smuggled my books to bed. Covered by the duvet, so that my parents would not notice, I read with the help of a torch. The idea of evolution, that all living beings had evolved from primitive organisms, inspired me to take up a study on comparative anatomy.

With my savings I bought some birds and fish from the fishmonger, and dissected them on my mother's kitchen table.

Some friends helped me to get hold of some dead cats, a dead monkey and even a porpoise, which was lying on the pavement outside a fishmonger's shop. It nearly made my mother cry out in despair. Having dissected the animal and cleaned its skeleton, I took my mother's biggest pot and boiled the oil out of the blubber, with the intention of using it for making soap. But, after a couple of days, I realised how upset and anxious my mother was. To relieve her of further worries, I poured all the oil – a couple of gallons of it – down the drain. That only made matters worse.

The collection of skulls and whole skeletons from this period became, in the end, quite impressive. Based on that, I wrote my own illustrated book on comparative vertebrate anatomy. It was a mixture of good observations and a lively imagination, but it impressed my biology teacher.

All this taught me a great deal, but my principal interest remained the life of the many animals I came across. I saw that animals fought for their existence in the same way as humans did. I often found that what happened amongst animals was cruel and heartless. To me they appeared to know love and hate just as we do. I came to regard it as certain that most animals have the same feelings as humans, that they can be loving and altruistic. This view, however, was modified with time.

The more I experienced, the greater my world became. I began to feel a degree of contempt for the many people who did not want to know that there are living beings as well as humans, who have a right to exist. The thought of people destroying and killing without conscience, worried me deeply. Of course, my own behaviour was not always consistent. I caught and killed mice and voles, and I found it exciting when a neighbour shot sparrows, but afterwards I always had a feeling of shame.

The emotions stirred by my contact with nature created in me an urge to explore and to spread a new gospel for the enrichment of mankind. I wanted to show people that their world was millions of times bigger than they had ever imagined. But how could a small boy ever hope to do that? It was important that I passed my school exams and was accepted into University. That was a hurdle which caused problems. I was definitely the hopeless blockhead of my class. During school hours, I often sat and gazed at the wall of the neighbouring building which blocked the view and did not permit even a glimpse of sunlight to enter the classrooom. As I sat there and looked at the dirty grey bricks, they took shape and became a green forest, with colourful parrots and monkeys swinging from branch to branch. And there were ferocious tigers too. It was difficult to penetrate, and I went stumbling over the creepers. It was so beautiful and exciting, all of it. A parrot nearby started shrieking 'Niels, will you repeat what I said!' – it was the teacher. The illusion had gone. I was back in the classroom and had to try to explain to the giggling class what the teacher had just told us.

School came to an end. That I ever passed my exams was a miracle. Later on I learned that it might not have happened without the help of some of my teachers, who understood that some very unusual interests and emotions had been disturbing my performance in all subjects except in nature studies. They had put up a heroic battle for my sake.

It was not only at school that I was a problem-child. My poor mother's patience was often put to severe tests. There was no end to what I might hide in my pockets – frogs, beetles, mice and many other goodies. It amazes me that she put up with my interests so well. The same, however, could not be said about certain other members of our family. One day, when I came home from a walk in the woods, my aunt, a very elegant woman, was having a cup of coffee with my mother. 'See what I've got,' I said, and pulled a grass snake out of my pocket. 'Isn't it lovely?' My aunt gave a penetrating scream, shot out of her chair and rushed out of the house. I ran after her to assure her that it was not a dangerous snake, but it was to no avail. She just ran, closely followed by me with my snake. This was the last time she visited my mother when I was at home. I only saw her again just before she died, many years later.

On another occasion, I came home with my breast-pocket filled with small yellow snails (**Succinia**) which I could not control. They were crawling all over my clean shirt. My mother was having a tea-party at the time; it is not difficult to imagine the kind of reaction her friends showed.

Although my interests were the cause of much inconvenience, I believe that they were shared, to a considerable extent, by my parents. Together we followed with enthusiasm the daily life of birds nesting in the Virginia creeper on our verandah. If a chick fell out of its nest, and we were not sure that its parents were looking after it, the whole family would care for it and feed it. On more than one occasion, I smuggled a nestling bird with me into bed at night because I feared that it was not warm enough. Usually it worked well, but I remember one morning waking up to find I had slept on a large blackbird chick. All that remained of it was a large pinkish-grey cake and a big stain on the mattress which my mother was unble to remove.

Every summer, we usually had the house full of young jackdaws I had picked up in the forest. They were delightful fellows who amused us all in spite of their incessant begging for food, their aptitude for teasing us, and their insistence that our best chairs were specially designed for them to sit on. Not many birds are comparable to the corvines with regard to intelligence. Among the birds I came to know well, only the grotesque ground hornbill seems to be their equal. Common to them all is their great attraction to shiny objects. They are terrible thieves. Numerous times I have seen them run away with their haul and try to hide or bury it.

The hornbill used to peck big holes in the ground to conceal his loot. But I do not know whether any of the birds ever returned to their hidden treasures.

Ground Hornbill

2.

My father died while I was sitting for my final exam at school, and my grandmother died a few months later. The result was that everything was in a turmoil when I started university. This, of course, affected my work badly. Moreover, I found it difficult to adjust to the independent study which was quite different from what I had been used to at school.

The University was, to begin with, a great disappointment. I found it hard to understand why I needed to learn chemistry. I could accept botany. The zoology taught seemed to be all anatomy, histology and classification. All we had to study was dead animals, or parts of them, well preserved in alcohol or formalin. No-one told us much about live animals, what they did or what they ate. It was not until a year later, when I got such excellent teachers as Lieberkind in zoology, and Hagerup in botany, that my study took a turn in the direction I had hoped for. The inspiration I received from these two men was invaluable.

At home, I began to experiment with bats. In the late evenings in the autumn, I could often hear the high-pitched call of the parti-coloured bats, but I never saw them. Their call was of so high a pitch that elderly people were unable to hear it. By experimenting a good deal, I discovered that it was possible to produce notes similar to those emitted by the bats. I only needed to dip a cork in oil and rub it against a sheet of glass. When I opened the window and rubbed the cork against the glass pane, I could entice the bats to come right up to me. Evidently, the sound I produced was similar to theirs; probably it was their mating call.

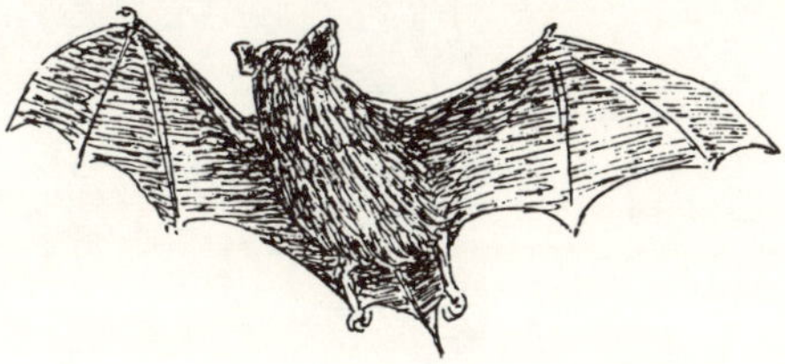

I told my teacher, Professor Adolph Jensen, about my discovery. He cleared his throat, but did not comment. I would like to study bats. How could I get hold of some animals? Adolph – so we called our professor – suggested that I should go round to a small back street called Pilestraede where they were demolishing some old houses. When I got over there, I was told by the workers that they had seen no bats. Couldn't I use bugs instead? There were plenty of those, but definitely no bats. What I had forgotten was that Adolph liked pulling our legs. When I told him about the bats, he must have been in one of his teasing moods.

At that time I was too much of a greenhorn to understand that it was not respectable to study animal behaviour. Such things could be left to the amateurs who had nothing else to do. The professional zoologist described the appearance of the animals and their structures and classified them. If interested in ecology, he could study the distribution of the animals and classify that. Also, their anatomical and physiological adaptations to the environment could be studied, but the study of animal behaviour was taboo. When I told Adolph about my unsuccessful visit to Pilestraede, he was inspired to further teasing.

A period followed in which he sent me climbing up one church steeple after another, until I had been to the top of most church steeples in the inner City of Copenhagen. I did not find a single bat. Later on, I was told that Adolph thought it fun to see how soon I would tire of the strenuous climbs.

Eventually I had to give up the study of bats, but before I gave up all hope, I was given a pair of noctule bats. Someone had found them when he cut down a hollow tree. It was late in the year and time for their winter sleep, so I placed them in an empty aquarium and left them there, being careful not to disturb them more than necessary. Next Spring, I gave them a rolled up mat to sleep in. The male turned out to be old and unable to fly. The female, on the other hand, was in excellent condition. Both animals had become very tame. Apart from a penetrating smell of urine from their sleeping quarters, they were a couple of charming pets. Every evening they came out to be fed with meal worms, scraped liver and cod-liver oil. This required great care. It was important that their wings and pelt were not fouled.

Unfortunately, I was still too inexperienced to be able to carry out any scientific work on my bats. I had no-one to advise me, and none of the technical aids used by later researchers. In spite of this, my time with them was very rewarding, and what they taught me about the handling of animals became very useful later on.

The two bats always returned to the same roosting place, and there was no need to confine them and restrict their freedom. They moved freely in my room. I always slept with my windows open, and there was nothing to prevent the female leaving if she so wished. It was not often she flew out of the window and when she did, she always returned at once, apparently not liking it outside in the street. There could be no doubt that the bats associated me with food and comfort. If at night they needed something, or their mat had fallen over, they always came and woke me up. The female swooped down over my head so I could feel her wings. As he could not fly, the male climbed up on my pillow so that I felt his soft pelt against my face. Neither animal would leave me in peace till I had discovered what they wanted and put matters right. Most often they wanted fresh water. When in July I had to travel to Hillerød to attend a course, the animals were so dependent on me that I could not leave them at home. The only solution was to put them in my pocket and take them with me. Soon I found that this caused no difficulties. I had hoped that my bats could sleep in the laboratory. This proved impossible because they liked to hide behind the cupboards, and my teacher, Professor Wesenberg-Lund, was afraid that they wouldn't come out. We therefore moved them down into the boat-house. They seemed quite happy there. Every evening when I went down onto the boat-bridge, they came out and were fed. Afterwards, the male always went straight back into the boat-house. The female circled a couple of times over the lake before she returned.

I began to feel that I was too tied down by my bats, and hoped they would leave me, but obviously they had no intention of doing so. Therefore, when I had to leave for Ireland, I gave them to a friend. Apparently they did not feel at home with him and disappeared soon after. I believe, however, that he managed to take a film with them to be used in schools.

3.

The bats were the last mammals I had for a good many years. My interests took a new turn. The cause of this was that I misread the question paper when I was sitting for my exam in botany, and consequently I did not answer the question set for me. When after the exams, I discovered my mistake, I went straight to my tutor, Professor Knud Jessen, and explained what had happened. He got my paper out, read it and said that it was excellent. I felt confident he would let me pass, but then he remained silent for a couple of minutes. Finally he said: 'One of the things an exam paper must show is that you can read your own language. Apparently you can't. Come back and sit the exam again.' I feel no doubt that Jessen knew what he was doing when he failed me. He was a kind and helpful man who allowed me to pick what flowers in the botanical garden I needed for coaching my fellow students. This I did in the laboratory. In the evening, Jessen sometimes called in to see how we were progressing, and to invite us all to his flat for a cup of coffee.

The extra time I now had to devote to the study of botany made it possible for me to penetrate deeper into the subject. Soon I discovered that there was nothing more beautiful than the delicate structure of flowers, and I began to regret my decision to become a zoologist.

One thing I had learned from Wesenberg-Lund was that bees and wasps are very interesting insects. I began to acquaint myself with the more detailed aspects of their lives. Wesenberg-Lund's interest in these insects was probably less scientific than his interest in freshwater fauna, but it was impossible not to feel enthusiastic when he talked about them. He used to scold me when I told him that I had destroyed a wasps' nest. Wasps were useful insects, he told me. He did not want to hear that they could also be harmful. Apart from such small arguments, I found that Wesenberg-Lund was a wonderful teacher. The memories of the hours we students sat around the table in his library listening to his accounts of the animals on which he was working, are among the most fascinating from my student years.

It was without doubt the combined influence of Knud Jessen and Wesenberg-Lund which finally made me decide to make bees and wasps, and the importance of these insects as pollinators of flowers, the subject of study for my Master's Degree. At that time, a shift in attitude towards animal behaviour was beginning to set in. This choice of subject for my special study was sufficiently respectable to be accepted by the new zoology professor, Ragnar Spärck.

The study of bees and wasps became a great joy to me. Whenever possible, I spent my time along the roadside or in a sand pit observing the insects. I was full of admiration for the energy with which the sand-wasps dug their holes and carried into them insects they had paralysed. It amused me to see the slim-waisted Ammophila run through the heather looking for a good big caterpillar to feed its larva with. Now and again, it made a buzzing noise as if it was humming a little song. Equally fascinating was to observe how precisely leaf-cutting bees gnawed sections out of rose leaves and joined them together into pretty little nests that resembled small cigars.

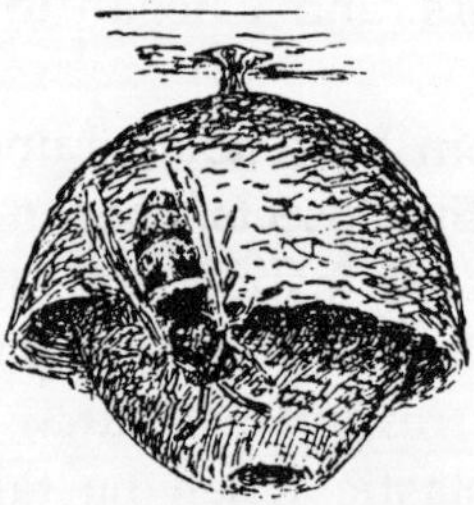

Wasp's nest

Many people thought, of course, that I had gone crazy when they saw me chase about with my insect net. This did not bother me. If only they knew what they missed. It always gave me a feeling of pride when I succeeded in stirring someone's interest.

In order to gain more experience, I went abroad during the holidays. I visited fellow entomologists and even rode my heavily laden bicycle round England and Ireland collecting insects.

One day, a friend of my mother's presented me with a little swarm of bees. This I placed in a small glass hive, which made it possible for me to observe what the bees were doing when at home. I placed the hive by the open window in the spare room in our flat in Copenhagen. I had read papers and books on the senses of insects about how they are guided to the right flowers. I had even read through Knoll's mammoth tome on "Insekten und Bluhmen". But most fascinating of all the papers I had read were those by Karl Von Frisch on the senses of the honey-bees and the way they communicate with one another.

I began to experiment myself. This resulted in my first scientific publication – how the contrasting colour patterns in flowers, those referred to as honey or, more correctly, nectar-guides, served to guide the insects to the nectar.

Later on, I acquired more bees and began bee farming in earnest, hoping to supply family and friends with honey. It was not a success. Exaggerated care almost killed the bees. However, during the war, when I needed hiding places for guns and ammunition, the hives became useful. Some of my bees were particularly vicious, and could be handled only by experienced bee-keepers. I came to the conclusion that what was hidden inside the hives would be reasonably safe.

I found the social wasps no less exciting than the bees. I found a way of habituating, as the psychologists would say, the wasps to my presence and the disturbances I might cause them. When I found a wasps' nest, I gave the branch under which it was hanging a good blow with a stick. Without changing speed or direction, I continued my walk as if nothing had happened. Of course, the wasps always came rushing out, but if I walked quietly on, they left me in peace. If I repeated this treatment several times during the day, the wasps soon became accustomed to it and stopped taking any notice. By the end of the day I could cut off the branch and, without any major risk, put the nest into a bag. At home I unpacked the nest and hung it up on my mother's balcony. Nobody was ever disturbed by the wasps and they never prevented my mother from having her small coffee parties on the balcony.

It is not nearly as difficult as many may think to observe what happens in a wasps' nest. All you need to do is to cut a hole in the side of the nest. It does not seem to worry the inhabitants too much. They are so accustomed to walking in and out of the opening they have made themselves, that it rarely dawns on them that they might take a short-cut through the new opening.

As soon as the wasps are accustomed to human presence, it is possible to sit quietly and enjoy the sight of life in the nest. There certainly isn't a dull moment. Every time a wasp returns with its prey – usually an insect larva or some other small animal – the whole nest becomes active. The wasps come streaming down towards the entrance opening. They push their way through the crowd, and those that come first grab hold of the prey and tear it with their mandibles until they have secured a part of it. Those that do not get quickly enough down to the entrance will do their best to steal bits from those that have already got something, and soon bits of the prey are distributed among many wasps. Stirred by the commotion, the wasp larvae become active. They move up and down in their cells like the pistons in a motor, a clear drop of saliva appears between the mandibles and collects on their almost bowl-shaped chest. Those wasps that have secured a morsel take it to a pumping larva and feed it. Then it licks up the liquid collected on the thorax. Sometimes, a wasp seems discontented with the amount of saliva it receives. It may take hold of the larva and shake and squeeze it until it produces one more drop. When a wasp, after feeding a larva, meets another wasp, an argument sometimes seems to take place. If the wasp it meets is bigger and stronger, the outcome will be that it regurgitates the liquid it has just received, and hands it over to its opponent.

Of course, I did not understand the significance of this, but later on, researchers discovered that the larval saliva contains something which promotes the development of the ovaries in the adult wasp. The more saliva a wasp receives, the larger its ovaries become. The larger its ovaries, the higher its rank in the wasp society – no wonder that wasps compete for saliva.

4.

During the war, while I was employed by the State Laboratory for Plant Pathology, I became fascinated by earwigs. These small insects, for which most people feel contempt, are not related to the wasps. I fell in love with them when I saw with what care they looked after and protected their eggs and young progeny.

Earwig in it's nest

Every earwig has a character of its own. Some were calm, others were nervous. Some were peaceful and others were bad tempered. These characteristics were particularly noticeable in their care of their offspring. Their sexual behaviour was, by and large, dull and lethargic. In talking to a small, private natural history society, which had asked me to give a lecture,

I gave a somewhat detailed description of the earwig's sexual life. This was apparently not what the society had wanted to hear and they were not pleased. Shortly after that the chairman 'phoned to ask me to give another lecture, but would I, please, be good enough not to discuss biology. I could talk, for instance, about the beauty of the butterflies. I was rather surprised by his request and pressed him to explain why I could not mention the biology of the animals. After some hesitation he told me that some of those that heard my lecture on earwigs had complained that I discussed a subject as delicate as their mating and copulation.

'I must confess' he added, 'that it also made me feel rather awkward when you mentioned it.'

When the time approaches for an earwig female to lay eggs, she digs a nest in the ground. She does not use her legs for digging as the wasps do. She digs with her mouth. One by one she loosens and picks up small lumps of earth and grains of sand with her mandibles and carries them away. She is in no hurry, but she does not waste time either. Without stopping to rest she walks quietly forwards and backwards with load after load. When the hole is deep enough she closes the entrance behind her with soil taken from the walls, thus producing a little chamber. When the little chamber is big enough she presses the walls until they are firm, and smoothes them with her abdomen. Now is the time for her to lay her eggs. She lays one egg at a time, takes a step backwards to pick it up and lick it. Then she puts it down in front of her. When the egg-laying is finished she has a tidy pile of 30-35 eggs in front of her.

During the next weeks the mother earwig guards her eggs. Incessantly she picks them up one by one and licks them. This, apparently, is important. If she is prevented from doing so or if she is removed from the next, the eggs will die.

A few weeks later the eggs have doubled in size. Through the transparent shell some of the features of the small earwig inside can be seen. During all this time the mother has had nothing to eat. When the young earwigs finally hatch out she is exhausted. But as long as she is capable she will do her utmost to defend them against intruders. The progeny remain for some time in the nest with their mother, but when soon after, she dies, she is devoured by her children, thus rendering them a last service.

5.

My time as a plant pathologist was no success. It was against my nature always to think about how to kill and destroy pests. A member of the Farmers' Council – the official committee advising the Government – said one day: 'We are not interested in the life-history of the animals. All we ask is that you find out how best we can destroy them.'

I was unable to understand how a responsible man could say anything like that. Why was it so impossible to make people understand that there are probably more useful insects than harmful ones, and relatively few insects interfere directly with our well being and existence. Were people really so stupid that they could not understand that unless we knew something about the animal's biology, we would never be able to give advice about their control. It worried me that we often killed many more useful than harmful organisms, and that any interference with the balance of nature might have serious consequences at a later date. The thought often pained me that the control measures we applied might cause severe sufferings to countless animals.

The difficulties I had with accepting my role as exterminator of harmful animals made me unsuited to my job. I wanted to know more about the animals and about their worlds, and I found it impossible to make any progress. In the end I succeeded – partly in secrecy – to complete a study of the senses and sense organs of fly larvae, and, consequently was awarded a doctor's degree. The success was the result of furious determination not to let anything prevent me from achieving what I wanted, or from reaching my goal.

Although I felt unhappy in my job, I stayed on for six years. There were two reasons for this; first, there were no vacant jobs to apply for; and second, it was war-time and our laboratory had become an important centre for anti-German activities.

Just like so many other young people, I was eager to do something to rid my country of the German occupying forces. This was a fight for human rights and I was naive enough to believe that everything would be well as soon as we got rid of the Nazis.

As soon as the war came to an end, I resigned from my job, hoping to go to England as a representative for Danish cultural life. Soon, however, I was to discover that my hopes for a better world were not as easily accomplished as I had thought. It upset me to see how revengeful my compatriots were. Nobody wanted even to try to understand the plight of the many German refugees fleeing from the battle-fields and temporarily housed in camps or school-buildings. Instead I heard hateful propaganda which more often than not was untrue. Intoxicated by their regained liberty, people forgot all about human rights and human dignity.

Disgusted and disappointed, I began to look for work outside Denmark. It was not long before an offer for a senior lectureship at the University of the Witwatersrand in Johannesburg arrived in the post. Shortly afterwards, I was on my way, together with my wife and our child, to a world about which I knew very little.

AFRICA

Our early days in South Africa were far from easy. Many immigrants gave up after a short time, but we were determined not to be beaten, and in the end success came our way. After a couple of years we had a small circle of friends, and with some money my wife had, we bought a half-acre plot in a blue gum (Eucalyptus) wood north west of Johannesburg. We managed to get a cheap loan from the Government so that we could build our own house. The plans were drawn up by a promising student of architecture at the University – the later well known Pancho Guedes.

Perhaps life in the woods was rather lonely – at least my wife thought so, but there were many compensations and we loved the place.

It was not long before we began to keep all kinds of pets. Apart from the usual kind of pets, such as dogs and cats, we kept more unusual ones too, and it was interesting to see how well they all got on together. They rarely did one another any harm in the house or outside. There could, of course, be disagreements when a new animal was added to the flock, but they usually settled in without too much fuss. My youngest daughter was presented with a pair of bantam fowls. Soon the hen arrived in our back yard with a batch of newly hatched chickens. There she met the cat whom she knew but always avoided. The cat seemed to regard the chick as easy prey. His intentions were only too obvious. Apparently he did not consider any possibility of unpleasantness.

Quite unexpectedly, the hen launched a violent attack. She flew right into his face and gave him such a hiding that he could do nothing but retreat. He never as much as looked at a small chicken again. Many a time I was intrigued when our cats walked across the yard, taking no notice either of hens and their chickens or of the flocks of weaver birds collecting crumbs outside the kitchen door. The animals knew and respected one another and did not bother each other.

The phenomenon of peace and neutrality around the home base seems common in nature. Doves and small birds are known to build their nests among the materials of eagles' nests, and they seem to be safe within a certain distance from their homes.

Rover, our big dog, never did any harm to our other animals. He was used to our guinea-fowls, for instance, and paid them no attention. One night a guinea-fowl fell down from its roost on our roof. I have no idea why that happened. We heard it fall and fly out into the woods. Next morning at feeding time, it did not turn up. The spoors showed clearly enough what had happened. Rover had killed it and eaten it on the spot where I found it. Contrary to what was usually the case, he showed no signs of guilt. I did not scold him because it had happened outside the area where peace had been declared.

Among our many animals I had a particularly soft spot for the donkeys. Ever since the days before the war when I was cycling round Ireland I felt attracted by these animals and wished to have one myself. My elder daughter had become very ill with Queensland fever, and I decided then to give her a donkey to ride when she got better. I bought Molly, as we named the donkey, from a farmer who lived about 30 km away. I offered to pay him forty shillings for the animal, which he refused to accept, but said if I would give him thirty-five shillings he would agree to the deal! Having concluded the transaction I was faced with the problem of transport. I asked the farmer whether he could bring the animal, but he did not fancy the idea. He thought I might get it into my own small car.

'Impossible!' I answered.

'Let us try', said the farmer, and so we did. He pushed from behind and I pulled at the donkey's head. Up it came. There was not much room. Molly could not stretch her back, but neither could she kick. As a matter of fact, she could not move at all. We got home without difficulty. I still remember the expression of amazement of some of the people we passed on our way who happened to notice my passenger.

Molly was a lovely animal, but she flatly refused to work. She preferred to have fun in the company of the other donkeys in the district. It was not long before we saw the result of this, and she gave birth to a daughter.

I had bought Molly so that the children could ride her, but this did not go too well. At times she would gallop quite out of control and push the children off her back by running under some low branches. Another trick of hers was to stop suddenly – the children would fly over her head and have a rough landing on the ground. Nobody blamed the children for giving up on Molly so soon.

My next attempt was to make a trap in which the children could ride. I made it from a box placed on an iron frame and a couple of bicycle wheels. When finished, it looked quite smart, almost as if it had been made by a professional cart-maker. The beauty of it was that I could also use it for transporting rocks for a garden wall I was building. But again, it was not long before I gave up using Molly for pulling the cart. It was much easier when I pulled it myself. She refused to pull the heavy load unless I gave her a good hammering, but I did not have the heart to do that. We therefore changed roles, Molly and I; I pulled the cart and she walked behind.

Molly and her friends had a territory where no other donkeys were allowed. Only at their drinking place did their territory overlap. But the two herds never drank at the same time.

One day, I harnessed Molly and drove over to a neighbour in the trap. He lived just the other side of a small brook which formed the boundary of our donkeys' territory. The grass along the brook was green and fresh, so I thought it unnecessary to tether Molly while I called at the house. I felt sure she would stay put and eat her fill. Hardly had I entered the neighbour's door before their maid called to me, Molly had disappeared. On my way home I found bits and pieces of the cart. It had been splintered against the trees. When I reached our house, Molly was grazing peacefully on the lawn. I never had another ride with her.

I have often regretted that I did not venture into a more detailed study of donkey behaviour. Unfortunately, such a study makes great demands on one's time, and I felt it difficult enough at that time to find time for all I wanted to do. It was a pity, all the same, because I had access to a herd of animals which were virtually living in a wild state, and yet they were tame and could be handled. Most of the time they were peaceful, but now and again the jack-asses fought one another. Sometimes these fights appeared to be violent. They kicked one another and seemed to attempt to bite each other's throats, but I never saw one of them actually injured.

When a jenny was to give birth, she always left the herd, and it was not until the foal was some weeks old that she rejoined it. To begin with, she had to be on her guard and protect her child against attacks from other donkeys.

It was a joy to us all to see Molly's foal grow up. Just like the infants of other mamals, foals love to play. Jane, as we named the foal, would gallop about and buck, and kick, and by doing so she probably learned to control her movements. A young foal can be dangerous until it has learned control and must be handled with care. Later on it can kick without causing harm. Many a time I was kicked so lightly by the bucking Jane that I hardly felt it. It amounted to no more than a light touch. I must, however, admit that I never stopped feeling anxious when she or another foal came galloping and kicking towards a guest or towards myself.

On Sundays I usually went out for walks with as many animals as wanted to come. Often I wondered what any stranger who might meet our small party would think about us. The party usually included the donkeys, our three dogs, a cat and a suricate – a kind of social mongoose. The latter was too short-legged to walk where the grass was long, and could only progress by jumping. Its small squeaking, complaining voice could be heard incessantly.

Having a somewhat sweet tooth, I often had some barley sugar in my pocket. Every evening on my way home from the bus, down across the open veld to our wood, I used to call the donkeys. If they were near enough to hear my voice they would come galloping. We then shared the contents of my pockets and walked home together.

It was not uncommon that an animal showed signs of jealousy. One day when out on one of my short walks with the animals, I probably paid a little too much attention to one of our pekingese dogs. Suddenly I saw Molly coming towards the dog, head down, ears back and teeth bared ready to attack. I was horrified, knowing that one good bite was enough to break the dog's spine, and I was too far away to interfere. But the dog apparently did not understand the danger. The donkey's posture was similar to that of a friendly, submissive dog. Therefore the dog stopped with its head high like a frightened donkey. The outcome was that the donkey only nibbled the hairs on the dog's back. To me, it looked as if the dog thought Molly was a nice, friendly donkey, and that she – the donkey – walked away believing that she had given her a good fright. This event was repeated a few weeks later.

Although we derived much joy from the donkeys, there were moments when they drove us to despair. It was almost impossible to keep them out of the garden. What made matters worse was that they invited all the other donkeys to join them. Fruit trees, vegetables, dahlias and other flowering plants were all targets for their appetite. If our donkeys were not satisfied with what the garden had to offer, they went into the house. They were particularly fond of home-made gingerbread. One day my wife went into the sitting room and found the donkeys devouring our full-length curtains. Although she afterwards did her best to repair them, she never managed to restore them to their original glory.

One evening, on my way down the hill from the bus, the donkeys did not come when I called them. They had not been near the house the whole day and they were not in the farmyard with the other donkeys of their herd. Nobody had seen them. We covered mile after mile in the car, but we never found them. Probably they had been stolen and sold to the horse butcher.

2.

As time passed, we gradually forgot the loss of our donkeys. We still had all the other animals. The dogs had puppies and cats, kittens. It was interesting that whenever an animal was to give birth it wanted our company. It even went so far that a kitten was born on my wife's shoulder. Afterwards I had the unpleasant job to destroy most of the kittens. It caused the children to regard me as a cruel father.

The kittens turned out to be the ideal pets for our two small girls. They were soft and alive. Much better than dolls. They were washed, treated with baby powder and dressed in dolls' clothes, and then given bottles and put to bed. Most amazing of all was that the kittens seemed to enjoy it.

Janet, our elder daughter, preferred the cats to all other animals. Our younger daughter, Kirsten, liked toads and the giant millipedes better. Whenever she saw her chance, she would smuggle them with her to bed. It was an often-repeated performance that we had to search her bed before we said 'goodnight' and turned off the light.

At that time we were visited by a journalist friend of ours from Denmark. One morning Kirsten came in to our friend who was still in bed. 'Aunty Esther, look what I've got!' she said, displaying a millipede reaching from her navel to her chin. Aunty Esther flew out of her bed with a scream. Poor Aunty Esther! We gave her a hard time.

Later on, in a book of hers, she referred with horror to the many evenings when we sat around the paraffin lamp – we had no electricity – and plucked ticks from the pelt of our dogs. Usually we put our catch onto the lamp, where those well engorged with blood exploded with a little pop.

Among other horrors were the solifuges, huge, up to two inches long, orange yellow, spider-like creatures which at night came in from the garden and, with an incredible speed, crossed the floor and ran up the walls. Although harmless, most people were afraid of them. We loved them and enjoyed seeing them catching insects and tearing them with their sinister looking nippers. Aunty Esther did not like them, but she controlled her fear quite well, although she could not help lifting her feet from the floor and uttering a faint 'Uh!' when one came running in through the open door and ran towards her chair.

We did not keep many reptiles as pets, although it was inevitable that we got a few. Our first one was "Waltzing Matilda", a female chameleon. It got its name because of its lurching gait. It spent most of its time on the curtains, waiting for flies to come within reach of its long, cylindrical tongue. Its big eyes, covered as they were by fused, scaly eye-lids, leaving only a small, round peep-hole between them, rolled slowly in all directions. It fascinated us that the eyes did not work together, except when the animal was taking aim at a fly. Otherwise they rolled independently in all directions, so that when one eye looked at the ceiling, the other might look at the floor, or out of the window. Since Matilda was our first pet from the wild, we were particularly intrigued by her. Apart from the entertainment she gave us, she offered us some protection, because many Africans believed that chameleons are possessed by evil spirits, an attribute they shared with the praying-mantis, which was known as the "Hottentot's God".

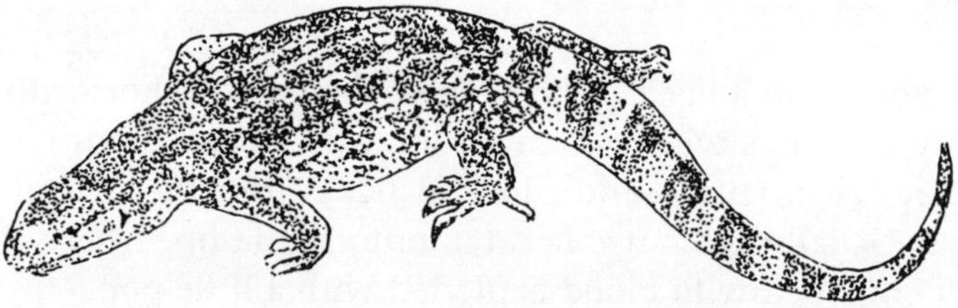

Tree Monitor

Once, my family and I spent a brief holiday on a farm belonging to the mother of one of my students. One day when we were out for a walk, I caught a monitor lizard, of a kind that spends much time in trees. It is a large lizard, related to the water monitor and the Komodo dragon. The animal feigned dead and it had its mouth wide open. I could not resist the temptation to point out to the student the many interesting features of the mouth, and pointed with my finger at the various parts. Without thinking what I was doing, I put my finger right into the open gape. Suddenly the jaws snapped together. There I was! I had walked into the trap and could not free myself. The student and I had to use all our strength to prize open the lizard's mouth wide enough for my finger to be released. It was bruised but otherwise unharmed.

Back in the University, I gave the monitor lizard into the care of Shadrach, who was in charge of our colony of laboratory rats. Together we decided to call the lizard Ben. Shadrach was, among the Africans, a highly respected medicine man, and the monitor lizard was the medicine man's sacred animal. My request that Shadrach should look after the animal therefore gave his authority a great boost. I told him that the lizard needed direct sunlight and not light that had passed through glass. He therefore took Ben with him on his daily strolls in the University park. This, of course, created some sensation. To begin with, he had Ben on a lead, but later on, when the animal was sufficiently tame, it walked freely close to Shadrach.

One day, when Shadrach stood in the open door with Ben on his shoulder a large lorry approached. When the black driver discovered the lizard he braked hard and quickly reversed the lorry to a distance he regarded as reasonably safe. Shadrach grinned but did not move from the spot. It was about ten minutes before the man plucked up enough courage to drive past. He left the University at a speed exceeding that permitted by the traffic regulations.

Although we sometimes kept the odd snake, none of us found them particularly interesting. Now and again we found some eggs under stones and brought them home. It was always exciting to see the small snakes hatch out. Only a few snakes were dangerous, and some of the less dangerous ones such as the herald snakes, we could handle with care. The children soon learned to recognise the smallish, fierce, spitting ringhals (ring necked cobra) and kept their distance. When frightened, this excitable snake ejects jets of poison through its fangs. It may cause blindness if it gets into the eyes, and can be dangerous up to a distance of six feet.

One morning, Janet found that one of her pet mice was missing. In its place she found a snake. It was stuck in one of the air vents of the box. Having devoured the mouse, it had become too fat to leave the same way it had entered. We let it out for Chippie, our suricate, to deal with. After all, she had been watching us with the greatest interest.

Chippie the Surricate

A DESERT INTERLUDE

In the mid 1950's there was an atmosphere of extreme mutual distrust between the national parks and the professional zoologists. They accused one another of hunting and shooting for pleasure, without regard for the well-being of the animal population or its ecology. Of course, nothing could be more erroneous. Two zoologists, the outstanding ecologist David Davis and I, saw this and we tried to mediate between the parties at various times. On one occasion when we had taken the ethologist Niko Tinbergen on a trip round the Kruger National Park, I went up to the Director for the South African National Parks, Rocco Knobel, to thank him for the facilities he had granted us. Knobel looked very nervous, and, before I could explain what I wanted, he began to protest that no one in the parks service would hunt, kill or interfere with any wild animal unless it was considered necessary. I knew he was right, but it was impossible to get a word in, and not until I prepared to take my leave did I succeed in saying what I wanted – that I had not come to criticise or accuse, but to thank him and the Board for their kindness towards us.

I don't know whether my visit to Knobel had anything to do with it, but some time later, the Board of the National Parks decided to organise an expedition to the Kalahari Gemsbok Reserve, under the leadership of Labuschagne, the Board's educational director. A number of other biologists were invited, and Davis and I were among them. Besides a group of zoologists and entomologists, the party included a vet, a botanist, a geologist and a palaeontologist, Dr Bob Brain, who is now a director of the Transvaal Museum in Pretoria. Together we formed an excellent team of experts, covering a wide field, and we worked well together.

Most of my research has been financed by myself. I have always been loath to ask for money when I could not guarantee results. I therefore designed all sorts of Heath-Robinson-like instruments with which I could record temperature, humidity, animal activity and other features of ecological interest. The instruments consisted mainly of old tobacco tins, bits of wire and string, bottles, an old alarm clock and sheets of paper. The physicists who saw it laughed and said it would never work, but it did work and, as I later found, it worked well.

The Kalahari Desert was a wonderful experience. I came to love the desert almost more than any other place on earth – the gentle colours, the teeming life, the peace. There were not only springbok, gemsbok, eland, lions, leopard, hyenas and other big animals, but also small animals, among which the suricates perhaps were particularly amusing. Most small animals are nocturnal. Among these were the spring hares which, when sitting on their haunches with their short front legs held close to their chin, always reminded me of the little old ladies of my childhood, sitting gossiping over their crochet.

Morning in Kalahari

The desert abounded with smaller rodents. Many of these never drank and existed on practically nothing but their metabolic water, topped up with the little moisture absorbed by seeds they had stored underground where the air is always slightly damp.

Outside in the open air, I found that the humidity fell below six per cent relative humidity. The kidneys of the mice had an astonishing capacity to concentrate their urine to a degree that measured several times that of humans. They only passed a small drop of urine outside their holes in the early morning. Otherwise they seemed to keep dry.

I took some mice with me to Johannesburg with the idea of learning more about their physiology. Among those I brought home was one species (**Desmodillus**) which at first I did not succeed in keeping alive. I could not understand why. The mice had plenty of dry food and water and yet they seemed to die from thirst.

Luckily I got the idea to add some green succulent food to their diet. As soon as I did this the mice recovered and thrived well. They were apparently behaviourally and not physiologically adapted to the desert. They had become dependent on the presence of green leaves and had forgotten that water was to drink. My observations showed that they detested getting their noses wet.

Not less fascinating were the black tok-tokie beetles (**Tenebrionidae**) who communicated their presence to their mates by tapping the ground with the tip of their abdomen. The beetles which were active by day had long spindly legs and were able to run at a considerable speed. Although these beetles did not leave their shelters in the hottest hours of the day when the surface of the sand got so hot that I could not kneel or put my hands on it without burning myself – nearly up to 70°C – they could be seen scurrying from bush to bush as long as the sand temperature was below 59°C.

Even that temperature is very high for a small beetle, and I began to wonder whether they were more tolerant to heat than other insects. Then I discovered that when the sand was hot they lifted their body up from the ground by as much as a whole centimetre. When measuring the temperature at that height, I found that the air could be 10°C cooler than the surface of the sand. In other words, the beetles brought their bodies up into a cooler climate. But even then, it was very hot, especially when one considered that the beetle's black body absorbed all the heat falling on it from the sun's rays.

I brought some beetles home with me to the laboratory in Johannesburg. Here, a few simple experiments showed that they were able actively to keep their body temperature below 40°C for shortish periods by ventilating their respiratory system and secreting a little fluid, thus increasing evaporation. It was interesting that a related white beetle from the Namib desert did not have this temperature regulation.

Although I concentrated my efforts on the small rodents and

the beetles, I was fascinated by everything; the birds, the suricates, the ground squirrels and what-have-you.

One lovely night, with the full moon shining over the dunes, giving the landscape an eerie appearance, I went out for a walk and decided to climb the highest dune in sight. I shall never forget my surprise when I reached the summit and saw a big fat frog (**Phyxicephalus natalensis**) sitting there. This was certainly the last animal I would have expected to encounter in the desert.

On a later expedition, I arrived at the desert a couple of days after it had rained. The desert was covered by yellow flowers and down in the dry river beds there were pools, or rather, patches of what looked like thick porridge, made up of mud and tadpoles. Two or three days later, the pools had dried up. Most tadpoles had died, but a few had succeeded in completing their development. They were now small frogs which could perpetuate the species.

On the last day of the first Kalahari expedition, I suggested to Labuschagne that he should ask all members to report on their findings before we broke camp. It turned out that all the reports, pieced together, gave a wonderful overall picture of the ecology of this particular desert. This inspired us to plan more expeditions to the Kalahari and to other places. Many years later Labuschagne told me that, with the first Kalahari expedition, we initiated the research into animal ecology which later on came to play such an important part in the work of the South African National Parks. For myself, the outcome was that the barrier between the Board and the zoologists was broken down and I was granted various facilities during my research visits to the Kruger National Park.

SMALL PREDATORS

One of the most amusing and endearing animals we ever had was Chippie, the female suricate.

The suricates are a kind of social mongoose. They attack snakes with no less fury than does their Indian cousin, the "Rikki-tikki-tavi" and other members of their family. Chippie always knew how to tackle a snake we gave her. She rushed, hair bristling, hither and thither, and made sideway jumps every time the snake lashed at her. Her ability to predict the snake's next move was fabulous. The snake seemed bewildered and showed signs of getting tired. Then Chippie lunged forward and, with a single bite, she crushed the vertebrae behind the snake's head.

Thereafter she immobilised the wildly lashing body by chewing its whole back thoroughly, from the head to the tip of the tail. Having finished this, it was time for the snake to be eaten. Gluttony is typical among suricates. Chippie would eat until she could not possibly get any more down, and all the while she was chatting to herself. Then, all of a sudden, she would roll over on her back and fall asleep. Her little belly was extended and round like a tennis ball. Only half the snake had been eaten. When, some hours later, she would wake up, the meal was resumed. She did not leave until she had consumed the whole snake.

The suricates are amusing animals to have around, because they are social animals which get very tame. Their little squeaking voices can be heard all the time and they seem incapable of shutting up. Some people may find this rather irritating.

Possessiveness is another characteristic typical to the suricates' make-up and this may easily lead to jealousy. I know of incidents when suricates actually chased visitors away. Luckily, our Chippie was better behaved. She liked to sit in front of the open fire and warm her tummy. When not doing so she insisted on sleeping in our best chair. A friend of ours once had a suricate which adopted a small chicken. It carried it to the fire and sat with it there when warming itself.

While travelling in Southern Africa I saw few animals which gave me as much pleasure as the suricates. They always became the centre of my attention when, in the early mornings in the Kalahari Desert, I observed the truly fantastic life stirring among the sparse vegetation of the sand-dunes. As soon as the copper-red sun was above the horizon, the suricates would begin to appear above ground. First came the adults and soon after, their young, almost a dozen of them. They would sit side by side in a row, all with their bellies towards the rising sun. Their backs were as straight as a ruler and their front paws were hanging by their sides. In their small faces the round eyes and the naked tip of their pointed noses shone like polished jet.

As soon as the suricates had taken their seats they were followed by the ground squirrels with their children. Often they came out of the same holes as the suricates and they were just as numerous. They too sat in rows facing the sun, sometimes next to the suricates. It always amazed me to see the predatory suricate and the ground squirrel living so apparently harmoniously together in the same burrows. An examination of the droppings of the suricates shows, however, that in the wild their principal diet consists of beetles, grasshoppers and other insects. Although they relish a good meal of a snake or a small rodent, the indications are that they eat vertebrates less often. Moreover, the burrows the suricates occupy are often dug by the squirrels, so why not leave them in peace to do the work?

There are in Africa various species of the mongoose family, but few of them are suitable as pets. It is only the social species that are reasonably tameable. The solitary ones cannot be relied on, and their bites can be serious. Only the banded and the dwarf mongoose and a few others seem to become good pets. When in Nigeria, I attempted to keep civets, but I soon found that these large animals were irritable and inclined to bite. Besides, their repulsive smell made them unsuitable as pets.

The palm civet and especially the beautiful genet were even better pets than mongooses. The young genet is an extremely graceful and amusing animal but it may be difficult to get on with the adults. They have another drawback as pets; they sleep during the day and wake up at night and want to play when it is our bedtime. My wife and I had many disturbed nights because our genet galloped backwards and forwards across the bed or even decided that it was more fun to play under the sheets.

Genet

Contrary to the true mongooses the genet has sharp claws which it can retract. They are excellent climbers and spend much time in trees. This created some problems in our household because our genet had an insatiable desire to climb the curtains. In the end I found this intolerable and one day, when it had an extra gay time, I decided to catch it. My attempt was far from successful. The genet flew at me with such a fury that I was forced to retreat. By doing this I had lost my authority and after that I could no longer handle the animal, but it remained as sweet and friendly as ever towards my wife.

In the end we succeeded in keeping our genet out of the bedroom. Being unable to spend the night with us, it began to stray. We saw it less and less. For a long time it came to be fed every night, but that also stopped. It had become a real wild animal, as it should be, and we rarely saw it thereafter.

2.

After many years in South Africa, we spent a couple of years in Uganda. One day a man came with a basket in which he had two small spotted kittens which had not yet opened their eyes. The man claimed that they were leopards but it was easy to see that they were something else – servals. I had difficulty in convincing him, and in reaching an agreement about the price, but in the end we made a deal and I became the owner of the most beautiful cats I ever had in my possession.

It took much time to hand-raise those two cats. It was necessary to check their stools daily and take care what they were given to eat. It was important that they had plenty of vitamin D, and they were therefore given great doses of cod liver oil.

It was lucky that the kittens were so young when we got them. Had they been much older they would probably never have become tame. As long as they were young they stayed around the house, where they had a good playmate in a young monkey. It gave us much joy to see them together, tumbling about on the lawn or chasing one another through the sitting room. Later on, when they began to stray away from the house, I felt it necessary to confine them to a large cage. They had no fear of humans and one could imagine what might happen if some native found them hunting their fowls. Wildfowl is the serval's natural prey.

Young Serval

The cage I built for the servals was a good roomy one, nearly 20 feet long and 6 feet wide, and next to it was a small shed where they could shelter. Their diet consisted of small birds which they consumed whole, feathers, bones and all. On days when we had no birds to give them they had meat, raw egg and cod-liver oil. To this we added hair, feathers or cotton wool. This addition was necessary to keep their digestion and motions in order. Many years later, I observed servals and other wild cats in zoos deteriorating and dying because they were not fed with whole animals, nor was roughage added to their meat. The argument was that it would shock the public if they saw them eat whole birds. People demanded that they were given prime meat. Offal looked too unappetising to please the visitors.

One day, an American zoologist came to visit us. He asked for my permission to enter the cage with the servals. I felt that I could not object to that, but because the animals liked to jump up on to people's shoulders, I warned him not to turn his back to them. After all, the cats were now as big as small dobermann pinchers, and I knew from my own experience that they used their claws to gain a foothold. They were fabulous jumpers. I had seen them jump on to a shelf six feet up from a distance of nearly ten feet without the slightest effort. A man's shoulder was not that high. Of course the man forgot my warning. He was so busy playing with the female that he did not notice the male. In one mighty jump it landed on his shoulder, slipped, and fell to the ground, taking with it the back of his shirt and a good deal of his skin, leaving deep bleeding scratches on his back.

During the following years, I became familiar with a number of big cats. There was the lovely American bob-cat, the lynx, leopards, lions, tigers and cheetahs. I loved them all, but none could measure up to the servals, and it was with sadness that I handed them over to my friend, the now well-known artist Jonathan Kingdon, when we left Uganda.

ZOO

Anyone who can get on well with animals should be thankful, for not everyone can. When dealing with animals, sympathy is needed. You must be able to identify with them and enter into a kind of comradeship with them, not attempting to make submissive slaves of them. If you wish to come to a deep understanding of them it is necessary to live with them and, as far as possible, share with them. Only a few people have the same luck as my family and I have had, to live together with animals as we did. Very few people have had a similar opportunity to develop their ability to handle and get on with animals.

The experience I had gained from my tame animals as well as from those I had observed in the wild became useful when, in 1962, we moved to Nigeria. At that time the University had a small collection of animals which were housed in an expensive but not altogether suitable building, and in open-air cages. Most of the animals were "presents" from people who had discovered that their pets did not always want to be petted, and took to biting – perhaps because their owners did not behave to their liking. Noticing how eager people were to be allowed to see a live antelope or monkey, I decided to ignore the wish of the University that I should dispose of the animal collection. Instead I would develop a real small zoo. My reason for this decision was that even in Africa there are very few places where it is possible to see wild animals. The animals would be useful, not only for our teaching; they could become a valuable tool in our propaganda for the preservation of the much too sparse wildlife. As I had seen the animals in their natural environment I no longer liked to see animals confined, but I regarded a zoo as an important and necessary means in our attempts to educate the public.

To begin with there was very little I could do to improve the cages. The threepence people paid to be allowed to see the animals did not help us much. Instead I improved the animals diet for there were instances of serious malnutrition. One day I saw a red-fronted gazelle stalk, capture and devour a rabbit, hair, hide and everything. Its daily food had been white bread, denuded of bran. Luckily we had no problem in getting fresh green fodder, and there were enough rabbits to spare as fodder for the few carnivores we had. When, to this, we added kitchen waste from the students' canteens, we could manage reasonably well.

Having improved on the feeding, I set out to train the keepers and to find jobs elsewhere for those I found unsuited to the work. The keepers should be able to observe the animals and to understand them. It was of no use that they took fright and retreated when threatened, thereby losing their authority. Authority had to be maintained as far as possible without resorting to force. By remaining calm and by confusing a threatening animal by acting differently from what it expected, it was often possible to avert an attack. 'Should a small monkey threaten to bite you' I said, 'give it your hand to bite. It will be so surprised that it will forget to bite hard, and perhaps it will accept your action as an invitation to play. If the monkey has developed its long canines, this may be a little too dangerous, but then you can push a carrot or a banana down its throat. The effect can be quite astounding. I must, however, admit that, faced with a furious baboon, I would beat a quick retreat'.

Not long after I had told this to the keepers, one of them forgot to lock no less than three doors leading to a cage with baboons. The old male in this cage was an extremely mean and dangerous animal, and we had already decided to have him destroyed. All we were waiting for was a date when the physiologists and anatomists could take him. When the old monkey noticed the open doors, it wasted no time in escaping. The keeper who left the doors open saw the animal on its way out and foolishly tried to grab it. Being no match for a 70 lb baboon, he was bitten so badly in the fingers that he later lost the use of them. Another keeper who tried to stop the fugitive was also bitten, though less severely.

From the Zoo, the baboon made its way to the Vice-Chancellor's garden where a chalet to accommodate important visitors was being built. When the labourers saw the animal they began to tease it and throw stones at it. The result was that the baboon turned on them. One man was so severely wounded in the leg that it took the hospital the greatest effort to save it. According to one doctor, they had "never before put so many stitches in one man."

Two days later, the baboon was shot in a forest area nearby. These events had nearly cost me my job, and my relationship with the Vice-Chancellor was for ever ruined. The event had, however, its amusing aspects. When the second keeper who tried to stop the baboon returned from treatment in the hospital I asked him how he was.

'It happened as you said' came the answer.

'What happened as I said?' I asked.

'Yes. It happened exactly as you said it would.'

I was at a loss. 'Listen. You must explain a little better what you mean. What did I say?'

'You said that, when threatened by a monkey, we should stuff a banana into its mouth. When I saw the baboon, I grabbed a hand of bananas. Every time it attacked, I pushed a banana into its mouth. It was not until I had run out of bananas that I had a bite on my finger.'

This was my worst experience as a zoo-man.

2.

It is very rare to see any wild animals in Nigeria, whether game or not. People there shot anything moving with their home-made muzzle-loaded guns, often without finding out what it was. Of course, the result was that they frequently shot one another, like the man I heard of, who thought he shot a monkey, and was surprised when his friend fell from the tree – stone dead.

In spite of this intensive shooting, there still seemed to be a small stock of animals which were worth hunting, and it often happened that a hunter offered me an antelope kid for sale. Usually they were kids of small duiker antelopes.

Duiker

On principle, I tried to refuse to buy animals from hunters, because I was against anything which could encourage the destruction of the sparse animal population, but every time I did so, the knowledge that the animal would die a miserable death plagued my conscience for days. The result was that, in spite of all good intentions, I acquired a number of antelopes which all eventually ended up in the zoo.

Many people who have attempted to hand-raise antelopes have met with failure. Antelopes are not the easiest animals to rear. They require the greatest care regarding cleanliness, food and bowel-evacuation. Most of them have difficulties in digesting cow's milk, which they must get used to. An upset digestion is the most frequent cause of death. The symptoms may vary, but I have seen more animals die from constipation than from diarrhoea. The cause is usually that those looking after them forget to massage their abdomen and anal region. Constipation may occur even after the animals have learnt to tolerate their food.

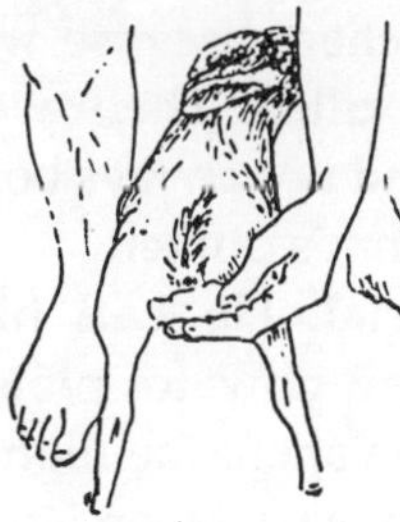

Stimulating the anal region

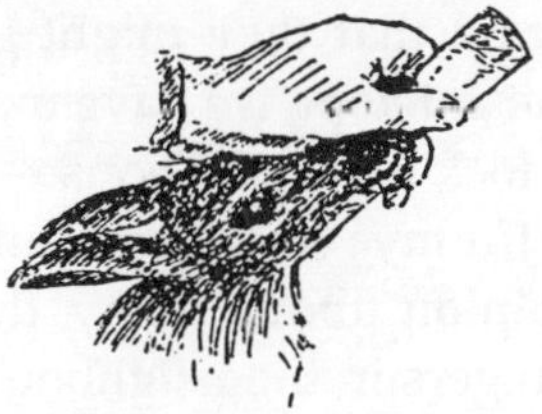

Hold the bottle this way

An antelope mother licks her kid for the first time when she cleans it after it has been born. In doing this she stimulates its intestinal movements. When the kid needs to evacuate, she licks it round its anus. Some kids simply cannot defecate without stimulation.

It is, of course, possible to spoil the child. It must learn to manage on its own. We had one duiker which we spoiled. Whenever it needed to empty its bowels it came to us and presented us its hindquarters. This behaviour did not cease until it was almost fully grown.

When it came to the supply of green fodder we had no difficulties, because our animals were allowed to roam freely and experiment to find out what suited their digestion. People who prefer to keep their animals in small pens should see to it that they give as wide a choice of plants as possible. Usually it is quite easy to obtain green fodder which the antelope will take, but it should be kept in mind that not all antelopes eat the same plants, and many do not care for grass. Most of them prefer leaves from shrubs or broad-leafed herbs. It is a peculiar belief held by many people that all ruminants feed on grass and hay, and also that bread is good for antelopes.

Antelopes are beautiful animals, and ours gave us much joy. It was particularly in the evenings, just before sunset, that they were most active. They could be seen galloping on our big lawn, wrestling or butting one another or playing hide-and-seek in the shrubbery. They could launch swift and unexpected attacks on their play-mates. When the game was over we shut them into a room behind the kitchen, because we were afraid that they might get stolen and killed. There was only one antelope we gave more freedom, and not always bothering to lock it into the house – Jimmy, a Grimm's duiker.

Jimmy was something quite special. He had his own opinion about many things and knew how to avoid most dangers in the neighbourhood. He often spent the night in the dense bush close to the house. We felt uneasy when he did not come home, and tried to look for him, but he was so well camouflaged that we could walk past him without seeing him. Most evenings he spent in the house, where he had his place on the sofa next to me when I was reading.

It was lucky that my wife did not object to Jimmy, as did the wife of a man I knew in East Africa. He had a bongo, a much bigger antelope than our duiker, which monopolised their sofa. She divorced him when she could bear it no longer.

Grimm's Duiker

Those lovely big, dark eyes so often seen in antelopes have led many people to believe that they are gentle animals. Nothing could be more erroneous. Even the small antelopes can be extremely dangerous, as one of our neighbours found out. Una Lister, the Professor of Obstetrics, had a nice, tame Grimm's duiker which, it was decided should be moved to the Zoo, an arrangement which caused no problems. Unfortunately, Una forgot to remove its beautiful tartan collar. One day when she visited "Dikkie" in the Zoo she decided to take the collar off him. Dikkie received her very well, but when she tried to undo the collar he protested and attacked, tearing her leg and hand with his needle-sharp horns. Nobody saw it, and Una felt ashamed. Although she was bleeding, she climbed the high barbed-wire fence which surrounded the Zoo. On her way home she met a colleague from the hospital, who gave her some assistance. It was with the greatest difficulty that he persuaded her to go to the Casualty department to be stitched up.

It cannot be emphasised too strongly that antelopes can be

very dangerous pets. The handsome bushbuck has probably accounted for more human lives than any other African animal.

Bushbuck

3.

Una Lister is one of the kindest and most warm-hearted persons I ever knew. Her desire to help did not only come to expression in her dealings with her many patients, but also towards the many animals people brought to her. When visiting her, one never knew what to expect. Her menagerie included not only the bull-terrier Charlie, who had sired most of the dogs on the University campus, but also numerous cats and a dozen grey parrots, flying freely in and out of her house, and nesting in boxes under the window. For a long time she shared her bedroom with a mongoose and she was deeply grieved when it died. Her ability to handle all kinds of animals was unique. When I got more animals than I could handle myself, I always knew that I could rely on her help. It was as a result of this that I gave "Piggie" the warthog, into her care.

Piggie was only a small pig when he arrived. He soon became very tame and moved freely in and out of Una's house. He became a great friend of Charlie and the two played happily together in the road. Apparently he believed he was a dog. His favourite place in the house was the sofa.

When Una had visitors who wanted to occupy the sofa,

Piggie demanded a place on their lap and expected that they would scratch him. Apart from the brush-tailed porcupine I once tried to keep as a pet, I have never come across a more uncomfortable animal to have on my lap. In the hot climate, I was always wearing shorts, which gave me no protection against his stiff bristles and sharp hooves. A certain hostility always reigned between Piggie and the parrots because he teased them and liked to bite their red tail feathers.

Piggie and the parrots did, however, share the same interest in things they could tear to shreds. Whenever we left a book for a few moments on the garden chair, we could be certain that when we returned a moment later, Piggie and Polly, the oldest parrot, had co-operated perfectly in dissecting it.

Even Una found in the end that there was a limit to what she could tolerate. We took Piggie to the Zoo and gave him his own sty next to the old warthogs. Piggie was very unhappy, he wanted to get out. In despair he jumped over the five foot high wall separating him and the old boar, who immediately stabbed him to death with his long, sharp tusks. Never had I heard of a pig jumping over so high a wall.

"Honk" a spurwing gander was another great character. He had reigned absolutely over the other animals behind the wire fence in Una's garden. Spurwings are the most aggressive geese I have ever known. When attacking, they beat at you with their wings, tearing you with a sharp claw on their thumb.

When Una had been granted leave by the University I offered to take Honk into the Zoo where he could roam freely. This, however, did not suit Honk and he refused to come when called. We tried to catch him, but this turned out to be more difficult than anticipated. He was remarkably cunning and avoided us skilfully. In the end Una succeeded in driving him into the fence, where I could grab him by the neck and put him into a sack.

Honk, with whom I had always been on very friendly terms, never forgave me this assault on his dignity. From that day on, I always had to be on my guard not to be attacked when visiting the zoo.

One day, going through the zoo, I saw Honk about a hundred yards away. He did not seem to have noticed me, so I hurried to the other end of the zoo on my errand to the baboon enclosure. Relieved that I had escaped any encounters with Honk, I got on with my work. But when, a moment later, I bent down to see better what one of the baboons was doing, I felt a violent blow from behind which made me lose my balance and fall, head first, into the fence, thereby upsetting the baboons and infuriating their old male. I cannot remember having had a worse hiding than the one Honk gave me. He must have followed me all the way through the zoo.

When Una later on moved to Zaria, in northern Nigeria, she took with her all her beloved animals. There she added more animals to her menagerie. Among these were a horse and a camel, a present from a grateful patient. She had it for many years until it died from indigestion, having eaten her servant's trousers.

4.

There is something about camels which I cannot learn to appreciate. I always feel uneasy when I stand in front of one of these animals, with its relatively small head on a long neck, which, like a giant snake, curves upwards from its clumsy, ungainly body. There is something menacing about its way of looking at you from under its heavy eye-lids. They give me the impression that they are both overbearing and insolent. There is also something slightly evil about the yellow teeth showing behind the cleft upper lip. I have often heard tales about people being severely bitten by malicious camels. Having said that, I must confess that I have never had cause to complain about the behaviour of the camels I have known.

In our zoo we had a camel which was very docile, and therefore excellent when I wanted to show my students its adaptation to life in the desert. Camels have thick callosities which make it possible for them to walk and lie down on the hot sand. They can close their nostrils when the sand is drifting, and they can manipulate the dry plant material they feed on without getting too much grit between their teeth. I liked to show all these things to my classes.

But our camel was useful not only for teaching. We also used it to transport green fodder from the forest. As a daily routine, a keeper took it out into the forest and there loaded it with branches and leaves. It happened, from time to time, that the keeper felt hot and tired, and thought he could steal a few moments' sleep when nobody saw it. The result was always the same. The camel would take a day off! Sometimes we had to search for hours, and on several occasions it was not found until late in the afternoon, several miles from the zoo. Such recurrent happenings were most annoying, but however much we scolded and threatened the keepers, we never managed to make them behave otherwise.

One day a new vet turned up. He was a nice young Canadian in beautiful white shirt, white shorts, white socks and white tennis shoes. He looked much too clean for his work. It was his first experience as a zoo vet, and he was eager to begin.

Not many days after his arrival, he came into my office and asked if I had any objection to him taking a blood-sample from the camel. Camels' blood differs from that of other mammals in having oval and not circular red blood corpuscles. Of course, I could not refuse him permission to take a few drops, I did not visualise that the camel would cause him any trouble if he gave it a small jab with a sharp scalpel, and smeared a drop of blood onto a glass slide. But this was not what the vet wanted; he wanted several cubic centimetres of blood.

'How do you think you can get that much blood?' I asked.

'That is quite easy. I can tap it from the great vein in its neck.'

'Do you think you can find it?' I thought about the felty mat of wool on the animal's neck, and its peculiar angular appearance. Nothing I could think of indicated the exact position of the vein.

'All right,' I said, 'it will be at your own risk. If you really think you can do it I cannot see any objections. But do not come to me and complain if things go wrong.'

Next day the vet came with all his paraphernalia, glass, syringes and whatever else is required for such an operation. As usual, he was immaculately dressed in white. It soon became obvious that it was less easy to tap the camel for blood than he had expected. He poked and poked with his needle into the protesting animal without getting a single drop of blood. In the end the camel did what is customary for camels when they cannot defend themselves by other means. It bent its neck and lowered its head over the man, and proceeded to empty its entire stomach contents over him. A thick, green, slimy soup enveloped him from head to foot.

Nobody has since dared to tap blood from this camel. It

took several weeks before it trusted anybody, and for a long time, it did not like anybody to come too close.

5.

There are two African animals which both seem completely devoid of any form of aggression. These are the pangolins and the ant bear or aardvark.

The pangolin looks rather like a cross between a giant spruce cone and a lizard. It was therefore not in the least surprising that many people found it hard to believe me when I told them that it was a mammal.

Pangolin at an ant's nest

Of course I wanted to know more about this animal and I managed to acquire a few specimens. Some of them were females with young ones. Usually they gave birth to only one young at a time, but one seemed to have twins. As they were of different size and seemed to be of different age, I concluded that one of them was an adopted child. The young ones just clung to her wherever she went and usually rode on her back. When they got hungry, they just climbed down to her belly to take a sip of milk from her nipple, which, among the pangolins, is situated in the arm pit.

The adult pangolin feeds on ants, which it licks up with its long worm-like tongue. To feed the animals I found it necessary to collect ants' nests from the trees in the neighbourhood daily and sustain numerous bites from these aggressive insects.

The pangolin species I tried to keep captive is an excellent climber which spends much of its time in trees. It has strong claws, well suited to its partly arboreal life and for tearing ants' nests apart. It also has a strong tail, with a very sensitive pad near its tip. When climbing, it supports itself with its tail, and may hold on to branches with it. It happened quite often that a pangolin which had climbed up above the door fell down and hit me when I entered their room. After some time I began to tire of them and when, one day, I found the burglar proofing, which consisted of strong, welded iron-mesh, torn to bits, I decided to set them free.

I have never had an aardvark myself, but we had a pair in the Oklahoma City Zoo in America, and my daily experience with them was such that I came to feel considerable affection for them. They are almost as large as domestic pigs and incredibly powerful. Although they could easily tear a man to bits with their strong hoof like claws, they were as tame and gentle as could be. I just could not resist their big, dark eyes, with their long eyelashes. Erroneously, we regarded the animals as slow and clumsy. One day, when their room was being cleaned, one of the animals escaped. It galloped through the zoo, soon out-distancing the pursuing keeper. The chase had a sad and abrupt ending when the aardvark landed in the lion's pit.

Aardvark

On my journeys in the dry parts of Africa, I have often come across the large holes in the ground dug by the aardvark, but I have never seen the animals themselves. They are very timid, nocturnal animals, which disappear down into their tunnels as soon as they notice someone approaching. To people travelling by night, their underground tunnels can present a danger. Once in the Kalahari Desert, the landrover in which I was travelling fell into an aardvark hole. On another occasion, when I crept down an aardvark tunnel, I discovered that a leopard had taken possession of it and used it as its shelter. I was lucky that the leopard was not at home.

6.

In Nigeria we often kept the giant pouched rats. They got as tame as guinea pigs and were easy to feed. We used them when teaching school children how to handle animals. They are also good for eating, and in Africa, where there is often a shortage of protein, many regard them as a delicacy. For this reason, I was eager to promote research into their domestication.

Quite often I was offered a young brush-tailed porcupine for sale. I found that they got very tame and were easy to handle. They are good for eating, but they are less easy to breed than the rats, and I gave up having them as pets when they began to gnaw the legs of the furniture. I decided then that they were more suitable for the zoo than for my home.

7.

While attached to the Oklahoma City Zoo I became particularly fascinated by the prairie-dogs. These charming rodents live in large colonies on the open prairie. Each family has its own warren, and each entrance is surrounded by a low wall which prevents water from entering the tunnel during the heavy rains.

Prairie-dog in the entrance of it's warren

When visiting a colony on a summer's day, one cannot avoid being impressed by the activities unfolding everywhere. This is especially true after it has rained. Then the animals will be engaged in improving on their dwellings. Soil that has fallen in is kicked out, and more soil is gathered to be added to the wall protecting the entrance. It is first scraped in under the belly, and then sent flying with a good backwards kick with the hind legs. When a good pile of loose soil has heaped up behind an animal, it backs, and with its paws pushes the whole pile forwards towards the entrance. Here it is added to the already existing wall and rammed down firmly with its nose.

Prairie-dog at work

One day we got a young prairie-dog in the zoo. It was so young that it could not possibly have seen the adult prairie-dogs at work. We put it in a small bird's cage with a bare tin floor, and an old rag to sleep on.

When we had had the animal for a couple of months, and it was well on its way to becoming a full grown prairie-dog, it suddenly began to scratch frantically on the bare tin floor, as if trying to dig. Imaginary, non-existent soil was pushed under its belly and then kicked backwards. Having worked like this for a while, the prairie-dog backed and then pushed the non-existent soil up to the end of the cage, where it was rammed down with its nose. It went through the entire behaviour needed for the wall building, although there was not a grain of sand which could inspire the animal to do what it did. I therefore concluded that the digging and building of walls could not be regarded as a sign of the animal being in possession of intelligence. To me the word intelligence implies both insight into the situation and understanding, in this case, of the consequences of the actions. This prairie-dog obviously had no understanding of its own behaviour. The actions, which followed a fixed pattern, were an expression of the organisation of its central nervous system. The animal had been frustrated by having no external inducement to dig and build. Its own internal need had accumulated to such a degree that it had burst through all restraining barriers and the actions were set in motion *in vacuo*. It had to carry out the sequence of actions, regardless of their appropriateness.

A fixed, innate pattern of actions, such as that described here, has been termed *instinctive* behaviour. Unfortunately, the term has fallen into disrepute among most biologists, because it is often difficult or even impossible to know whether an act is innate or learned. There is a tendency to talk about instincts in a much too casual way, thereby covering up an inadequate knowledge.

8.

One morning, when I opened my post, I found a letter from eastern Nigeria asking if I could accept a tame coati female for the Zoo. Although it was not an African animal – coatis live in Central and South America – I was grateful for the offer, because we had no live representatives of the group of carnivores to which dogs, bears, coatis and weasels belong. The coati was just right for demonstrating to the students how a member of this group had been adapted for its peculiar mode of life. It was an arboreal insect and fruit eater.

The coati arrived by air in a small box. A bit cramped, I thought, and felt sorry for it. It seemed perfectly tame and loving, so I decided to let it loose in the car. All went well. Zotie – that was the name it had been given by its previous owner – behaved like a perfect lady, sitting in the passenger's seat next to me. We stopped by the Zoo gate and I wanted to carry Zotie to the cage made ready for her, but she refused to let me come near her. In fact, she did not want to get out of the car. In the end I got tired of the game, and tried to pick her up by force, an action she much resented. She flew at me like lightning, and dug her sharp canines as deeply into my thigh as they could go. I did, however, get a good grip on her, and carried her successfully to her cage before getting my wounds patched up.

Zotie, whom I soon forgave, was a spoilt, loving creature who always wanted someone to be near her and pet her. To give her company, we imported another pair of coatis, but those we got were unfortunately of a different species and Zotie would not have anything to do with them. She was orientated towards humans and wanted only human company.

One day I got into the cage to play with Zotie. She was as loving as ever but when I wanted to leave her she would not let me. Every time I approached the cage door she became a bristling bundle of fury, and barred my way. Remembering my painful experience on her day of arrival, I dared not assert myself. Over and over again she drove me into the furthest corner. I was completely stuck for nearly half an hour when a keeper finally appeared and rescued me.

9.

There is, of course, no reason to imagine that a hyena feels worried because no-one has anything good to say about it, but it saddened me. I felt upset when people talked about it as ugly, ungainly and cowardly. Nobody realised what an interesting and peculiar animal it was.

Young Hyena-cub

It must be admitted that the hyena does not conform with the human concept of beauty. Its weak hindquarter does not seem to fit with the powerful shoulders and heavy head with the dark melancholic eyes. Unfortunately, I never succeeded in carrying out a proper systematic study of them, and none of those scientists I have approached and tried to persuade to look at their anatomical and physiological adaptation have ever done so. Only Hans Kruck and a few others have worked on their behaviour. Although not much bigger than a dog, the hyena has a windpipe and lungs like an ox, and it has a peculiar adrenal tissue. Nobody seems interested in finding out why this is so. Probably the explanation for the scientists' attitude is that the unnatural laboratory rats are much easier to handle, much cheaper, and yield quicker results.

My own impression of the wild hyenas is that they are shy

towards man and get frightened easily when single and not in a pack. On the other hand, I once met a wild hyena which was so preoccupied with emptying a rubbish bin that it did not notice me standing less than six feet from it.

Many tales are told about hyenas having attacked people at night while they slept. I did not believe these stories until one night a hyena came without my noticing it and stole a buck skin lying by my side. The camp fire had not deterred it. When, in the morning, I woke up, the skin had gone, and the big footprints told their tale.

While in Nigeria I bought several young hyenas. They were peculiar timid animals which could be tamed. Because the female has a clitoris as big as the male's penis they are almost impossible to sex. When the female is on heat her clitoris opens with a slit so that copulation can take place. Both males and females have a false scrotum. The male's testicles remain up in the abdominal cavity and do not move down in the scrotum as in most other mammals.

Although the hyenas look like ungainly dogs, they have very little in common with them, and it is difficult to learn to understand them. Their behaviour is in many respects similar to that of the mongoose, to which they are in fact much more closely related. I tried to play with those I had, but that was difficult because I had to be on my guard not to lose a finger. Even young hyenas have terrifying jaws and teeth. I tried once to train a young hyena to go out for walks with me, but this attempt was not very successful – probably because I was not an experienced trainer. Of course I had it on a lead, but it would not walk nicely by my side. It pulled in all directions. It amused me that nobody seemed to realise that it was a much-feared wild animal I had by my side. Apparently people thought it was a big puppy.

In the end I gave up having hyenas at home. Their wild shrieks made the neighbours think I was beating my wife. This, and also their unpleasant smell, made me decide to abandon the experiment.

THE BIG CATS

To many people befriending a large animal has a romantic appeal especially if the animal is an unusual one. If it is potentially dangerous the attraction becomes even greater. It seems to inflate our ego and give us self-esteem. I must admit, that in Nigeria there were moments when I felt a certain satisfaction when I was standing with the elephants and patting them in front of an admiring crowd. I remember being asked what "ju-ju" (medicine) I was using to gain so much power over the animals. By and large, though, I do not think anyone can accuse me of ever having suffered badly from this "disease". Showmanship is not the same as love of the animals. I do not approve of the former, and I find it repellent when someone demonstrates his inferiority complex by appearing together with a wild animal or with a large python wound round his neck. I have loved many big animals with which I have had dealings, but I cannot truly say that a lion or a tiger is a better pet than a common cat.

Many Europeans settling in Africa have attempted to tame and to keep big cats as house pets. This is certainly possible, provided one can afford it. Their food is expensive, but otherwise they demand very little, and most of them are relatively easy to control. Nevertheless I always feel that the best place for big, captive cats is the zoo, where they are always a great attraction.

When, many years ago, I was planning an expansion to the Ibadan University Zoo, I uttered a casual remark at a committee meeting about how easy it is to confine lions. 'All you need,' I said, 'is a high fence similar to that around the tennis courts of the University.'

The Vice Chancellor was struck with horror. In his opinion, the main aim in life of every lion was to devour people, among whom they regarded Vice Chancellors as a particular delicacy. A committee was appointed there and then to investigate whether I was suitable in my job and knew what I was talking about. I was to be censured.

Two years later, when the Vice Chancellor was too occupied with politics to bother with anything else, Dr Una Lister and I each bought a lion cub. They were miserable specimens, suffering from hookworm infestation. My cub lived only for a few days, but Una's lived for another year before it too died from hookworm.

Una often went out for walks with her lion, but this did not always go too well when there were cows about. Cows held an irresistible attraction for the lion and Una was not strong enough to restrain it. The result was a wild chase – the lion after the panic-stricken cows and Una after the lion. This, of course, did not enhance her popularity.

It had been the intention that Una and I should present our lions to the Zoo. Now that both lions had died, and I had departed to America, the Zoo still remained without these carnivores. However, a couple of years later my helper, Bob Golding, did succeed in building an enclosure like the one I had suggested. At that time the Vice Chancellor had fled from the civil war and settled in America, and nobody raised any objections.

The Zoo acquired two lions, one from England and the other one from East Germany. Then, as now, there was a surplus of lions in European Zoos, so it was easier to get them from Europe than from Africa, where they are rare in most places and protected by the conservation laws.

It was a thrill to see the lions amidst the natural vegetation where they belonged and where they seemed to feel at ease. Soon they began to breed, and, during the years that followed, quite a number of cubs were born. To the public they became a great attraction. To me it was a joy to watch the mother care for her young, and the manner in which she saw to it that each one of them had its share of meat. She kept the male at bay until they had all eaten their fill.

It is possible to rear rats together with cats so that they

become play mates and live together in peace. The drawback about such a partnership is that the cat never quite realises its own strength, and may unintentionally hurt the rat. My own experience from playing with lion cubs is that I more often than not received numerous scratches and bruises. When a cat hurts its play mate – whether a rat or a human – it is not out of ill will. Accidents may easily happen, and it is understandable if the neighbours of those keeping the great predators as pets start to complain. Their owner must keep a close check on them. When they are not confined, but allowed to prowl about, this is more difficult than with dogs.

Once, as a friend of mine got out of his car outside a hotel half way between Mombasa and Nairobi, a fully grown lion jumped at him, placing a big paw on each of his shoulders. He fell flat on his back with the lion standing over him. He thought his last hour had come until the hotel owner appeared and, grinning, told him not to worry. It was a tame lion which would do him no harm. No-one could blame my friend for passing a few remarks which were anything but polite.

When the big cats are kept as house pets the relationship with neighbours and visitors is often strained, and there may be difficulties within the household itself. Once I met a store-keeper who had a tame lion, which he had got as a small cub. He and his wife thought they would have it sleeping with them on a mat next to their bed. This idea was not shared by the lion. It wanted to sleep in the bed. Every night it crept up into the bed and found a comfortable place between them. Lions grow to a considerable size, and in the end there was not room for all three of them. The wife therefore gave up her place to the lion and moved to another room. After some time, however, the store-keeper found that his pet caused him too much trouble altogether and got rid of it. Presumably, his wife came back.

Although lions rarely kill unless they are hungry or feel threatened, I do not like meeting them when I am out of the car. Young lions can be very inquisitive, and I have experienced that they have followed me through dense bush. I knew they were there, and from a distance my colleagues saw them, but they were hiding from me until I got into the car. Apparently they wanted to see what I was doing. On such occasions I cannot help feeling uneasy. It is curious that lions do not seem to associate cars with humans and it is possible to drive a car right up to them.

The wild lions I have met on my walks never seemed frightened, but neither have they appeared particularly courageous. Once, my big, strong Bantu guide and I came across a couple of young lions resting under a bush. We watched them for a while.

'Wait here,' said my guide. 'I will show you something.' Without further comment he ran towards the lions, swinging his arms in the air and roaring at the top of his voice. Both animals bolted like rabbits and disappeared in the long grass. It is, however, dangerous to conclude from such experience that lions are easily frightened, nor can one accept the old belief that lions are afraid of fire. This I learned when, one night, we camped next to an artificial pond where the wild animals used to come to drink. We had no tents, and most of us slept on the ground between the big camp fire and the cars. In the early part of the night, our sleep was disturbed by the roars of a lion nearby, but we took no notice, thinking the camp fire would keep him away. At four in the morning, my neighbour woke me up.

'The lion is here,' he whispered.

Sure enough, there was the biggest male lion I had ever seen, having a sip of water next to the camp fire, no more than a dozen paces away from where we were. He watched us for a few moments and then continued drinking. This done, he walked calmly away.

2.

One day in Nigeria I received a cable offering me a leopard for our University Zoo. Of course, I accepted. A few days later I got a message that the animal was on its way and that the plane bringing it could be expected to land at the airport any moment. On my arrival at the airport the plane had already landed and was parked outside the building. A small crowd of frightened and bewildered men stood and peeped into the open hold, where the leopard was in a small, flimsy wire cage. It was beyond me how it had been possible to get a big leopard into such a small cage, but there it was, and we had to get it home to the zoo. Later on I learned that it actually had caused the former owner some problems. He had grabbed the cat by the scruff of its neck while his father held on to the tail. Together they had then tried to get it into the cage. At their first attempt they failed miserably. The father was rather badly mauled and had to be treated in hospital. They succeeded at a second attempt. I was also told that the reason for offering the animal to me was that it had first attacked and badly mauled a native boy and, shortly after, attacked the owner's highly pregnant wife.

When I arrived at the airport I had no idea of the leopards past. When nobody else dared to approach the cage I climbed up into the hold. We could not leave the cage where it was. I had to get it down into the car and over to the zoo, where the big cage had been made ready. No-one could doubt that the animal was terrified. Every time I tried to approach, it hissed and lashed at me with a fury I had rarely witnessed before. It took much cajoleing and persuasion before a couple of the bravest men ventured to lend me a hand. We got the cage safely on to the Landrover, but it proved to be a more difficult job than I had anticipated, because the cage had no handles.

It did not take long for Cheeky, as we named the leopard, to settle in. She became very tame and loving. Whenever she saw me she followed me with her eyes, and I always felt that I had to put my arms round her neck and love her a little. It was hard to think that she was a dangerous animal. But dangerous she was. This she demonstrated a few months later when a small African boy got inside the safety fence. Unexpectedly she lashed out at him, luckily without getting hold of him. The horrified boy fled to safety, screaming: "Uh! Him likes smallchop too much." (He's too fond of small snacks.) Thereafter we put up a placard with a cartoon, and the words: "Him likes smallchop too much." Since then we had no more startling events, but on several occasions I noticed that Cheeky reacted differently to children than she did to adults. Children she undoubtedly associated with food.

Cheeky

Unfortunately, Cheeky did not live for very long. An undigested bone splinter got lodged in her rectum and she had to be operated on. With some difficulty, we managed to give her an injection. It failed to have the desired effect, so she was given a second one. This time she fell asleep. She never woke up again. At two o'clock in the morning I came to see how she was. I found the vet sitting with the leopard on its bed of straw. Tears ran down his cheeks. Cheeky was dying.

ELEPHANTS

There was a limit to how many animals – and how big they should be – I thought we could have in our small zoo. Our finances were very limited and it was difficult to find the money to cover our expenses. It had happened on several occasions before my arrival in Nigeria that a small elephant orphan had been handed in to the University. They all died within a few days. Nobody had ever succeeded in hand-rearing an elephant from birth.

One day I discussed with our staff which animals we could accommodate. I said categorically: 'One animal I definitely do not want in the Zoo is an elephant.' I could not possibly imagine how we could house an animal of that size. Even the problem of getting enough fodder was frightening enough.

A week later I had an urgent telephone call from a European ranger in Ilorin, a town about 90 miles north of Ibadan. He wanted to know whether I could accept an elephant for the Zoo.

My answer was: "Definitely not."

'It is only a small one; just something for you; I was forced to shoot its mother some time ago. She was causing too much damage. The young one is old enough to look after itself, but now that too is causing too much damage. We must get rid of it.'

'I can under no circumstances have an elephant here! How do you expect us to accommodate it? We have neither room nor enough money to keep such an animal.'

'It won't cost you anything. Don't you think you can find room for it?'

The voice began to take on a more pleading note. I was told that the elephant was only two or two-and-a-half years old. At the time when the mother was due to give birth, the herd to which she belonged was on its way north to the banks of the river Niger. She had left the herd in search of a place where to deliver her child in peace. The herd did not wait for her, and she had not been able to join up with it again. If I refused to take the calf, it would be necessary to shoot it. Could my conscience bear that?

This argument settled the matter. The thought of killing a young elephant was more than I could bear. Just like so many others who have experienced wild elephants, I felt the greatest admiration for this impressive big game, but also like so many others, I found it awe-inspiring. It was therefore with trepidation that I allowed myself to be persuaded to take this young elephant.

'When do you think you can bring the elephant?' I asked.

'I'm sorry, but I can't bring it. You'll have to collect it yourself. I have no transport.'

I thought for a moment. 'Nor have I, but perhaps we can borrow a lorry from the University.'

How on earth we could hope to transport an elephant on an open lorry was really beyond me, but perhaps I could find a way of solving the problem.

'Where's the elephant?' I asked.

'About eleven miles from here.'

'Where abouts?'

'I saw it there a couple of days ago near a village. It's a lovely youngster. I got right up to it, so close that I could almost touch it. It's almost tame.'

I sensed more problems.

'You mean to say that it's still free in the bush – that you haven't caught it yet?'

'That's right. You'll have to catch it yourself. But it shouldn't be too difficult. It isn't in the least frightened.'

The ranger promised to organise help from the local people if I would bring some experienced keepers and a vet to look after the animal on the way home. The whole situation was much worse than I had anticipated. I did not have the faintest idea how to tackle the situation, but the problem had to be solved. We cut down a lot of teak trees in a plantation belonging to the University and used the trunks for building a stockade around an area of almost an acre. Also, with the help of tree trunks, we constructed a small pen on the back of a lorry. Nine days later, everything was prepared, the expedition organised, and we were ready to start.

It was Ramadan, the Moslems' new-year, when we started. We spent the night in a small hotel in Ilorin. This was the town nearest to where the elephant was supposed to be. The streets in the little town were crowded with people. They had come, some walking, others on horseback, to be present at the festivities. There was to be a great show of horsemanship in the square in front of the Emir's Palace. Chiefs from all corners of the Emirate had arrived to pay their respects to their ruler. However, coming from the University, we were too busy to bother about the celebrations. We got hold of some young people from the club and persuaded them to join us in the hunt. At seven o'clock the next morning, our now reinforced expedition left for the hunt.

The area in which the elephant had last been seen was only a couple of miles from the main road. The inhabitants of the village nearby were summoned to a meeting. After brief negotiations we engaged the most fit and able men to help us. The remaining population was given strict instruction to remain in the village. We constructed a trap, made of strong ropes. A good friend of ours, the Danish Ambassador Børge Raavad, who was among those who had joined our party, was put in charge of the trap.

Unfortunately, we soon learned that the men we had engaged were absolutely useless. They completely ignored our instructions. They had been told to surround the elephant, but instead they approached it from one side only and chased it, while screaming at the top of their voices. It was out of the question that we would be able to drive it into the trap.

The general excitement was more than the people in the village could stand. Soon they forgot that they were supposed to stay at home. It was impossible to control the screaming hordes who now joined the hunt. The poor animal ran and ran, and we ran after it with ropes, tranquillisers and whatever else might be required.

The temperature rose as the sun climbed higher in the sky. We got dripping wet with sweat, but we had no time to relax. Having started the hunt, we would have to carry it through to the end. We tried to get a rope around the elephant's legs, but every time we tried, the noose slipped off again. A strong and solid English girl managed to grab hold of the elephant's tail. She was dragged some distance along the ground, but this did not slow down the elephant's speed. We began to get desperate. This was one of the most strenuous jobs I had ever undertaken.

Not until about three o'clock p.m were we successful in getting a noose round the elephant's right hind leg, which stayed on. Soon we got another one round its left leg. The vet gave our captive an injection of a tranquilliser and we secured it with more ropes. The triumphant march to the car began. The hysteria which now developed exceeded all my expectations, I found it revolting. The villagers could not leave the poor animal alone, and in the end we had to agree to taking it through the village so that everyone could see our catch. I tried to keep the crowd away from the animal, but it seemed a hopeless task. When I was threatened with a knife, I was urged by my people to make no further attempts.

Back at our cars, we were faced with the problem of getting the elephant up onto the lorry. The "small" animal showed no signs of having lost its fighting spirit. It charged desperately, and lashed out at us with its trunk. In the end, however, we managed to get it under control, and by constructing a ramp from some young tree trunks, we finally hauled it up and secured it well between wooden bars.

For courtesy reasons it was necessary to pay a visit to the Chief Secretary of the Emirate. He was a friendly man who treated us to some much-needed cold drinks, but we were unfortunately compelled to cut our visit short because the Emir 'phoned to say that he too desired to see our catch. This request worried me, because we would then have to drive through the throngs of people at the approach to the palace. I told the Emir so, but he insisted on us bringing the elephant to him. He promised us police protection and said that we would be quite safe. We could park our cars in the courtyard of the palace.

As we approached the square in front of the palace my worst fears came true. A screaming, roaring mob gathered around the lorry and tried to jump up on it. There was nobody to protect us. At last we reached the tower gate and were let through by two policemen. We parked the cars by the front door of the main building. My wife and I were conducted into the front room of the palace, where we were offered a seat on a pair of chairs of the kind often seen in the gardens of English country pubs. We were told to wait there until it suited His Royal Highness to receive us.

Out in the square, the noise rose. Everybody wanted to see the elephant. There was nothing the two policemen could do to keep the crowd at bay. People in the rear pushed on those in front, and in the end the pressure became too great for the flimsy lock in the tower gate. Soon the courtyard was filled with the excited mob.

The Emir was reclining on a chaise-longue covered with a blanket of red plush at the far end of the throne room. On the floor, along the wall on either side sat a row of dignitaries. Those paying homage were led in, one by one. They threw themselves on to the floor in front of the chaise-longue and muttered a few words. Then the Emir gave a brief reply and indicated with a movement of his head that the audience was over.

At long last the Emir got up from his repose and came out to see the elephant. As he appeared in the door an ear-splitting pandemonium broke loose. He smiled and waved his hand, apparently flattered by the unrestrained ovation.

"I want to see its feet" he demanded, after giving the animal a casual look, so we had to let down the backboard of the lorry. After another brief look, he said: "You may go" and returned to the throne room.

The situation in the palace-yard had meanwhile taken a serious turn. Everybody wanted to see and touch the elephant. The two policemen were apparently afraid of the crowd and did nothing. It was not until my wife and I began to hammer those nearest the lorry with our fists that they picked up some courage. Our efforts were wasted. There was nothing we could do to restore order.

As this hubbub mounted in a fast crescendo, a new crisis developed. The elephant collapsed. Whether the ropes were too tight, as suggested by my wife, or the strain and panic were too much, I cannot tell. The animal was lying on its side and it seemed to have stopped breathing. Quickly we loosened the ropes and I knelt down on its chest to administer artificial respiration. For a few agonising minutes, I pumped air in and out of its lungs. Then suddenly its eyelid began to quiver and its respiration began to work on its own. A moment later we had the elephant back on its feet.

But now we were faced with new difficulties. In the crush a number of people had fallen and were now lying, wounded or unconscious, on the ground. We sent a message to the Emir, asking for his cars to transport the casualties to hospital. Although he had quite a fleet of vehicles, he refused. Instead, he demanded that we should use our own car as an ambulance. My wife was graciously granted permission to sit on the garden chair in the ante-room while I made three or four trips to the hospital with the casualties.

At long last it seemed possible to leave for home. The sun was setting and darkness was quickly approaching. I called my men together. One of them was missing. To look for him was like looking for a needle in a haystack. I climbed the gate tower and looked from its roof at the crowd. It did not look too promising, but if I could not spot him, perhaps he could see me. Just as I had given up all hope and decided to leave without him, I found his grinning face below me.

It was a little past midnight when we arrived in Ibadan. We had no difficulty unloading the elephant and getting it into its pen. Soon after, it slept peacefully with its trunk on my lap.

We always named our animals. It was my idea that the elephant should be given a real Nigerian name, but the keepers were unanimous that we should call it Dora, so I agreed. This, however, did not please the newspapers. They protested: "Why not give a Nigerian elephant a Nigerian name?" wrote one of the papers. But now it had been named, the keepers found it improper to make any changes. It therefore kept its name for the rest of its short life. It died about ten years later, when a vet gave it too big a dose of narcotics. He wanted to examine one of the tusks, which he thought might need some attention.

Dora was a gentle and peaceful animal. Only once, when she was frightened by two young baboons that had escaped, did she attack me. I was in the pen with her, trying to calm her down, when she went straight for me. I was lucky to catch hold of her trunk and dig my thumbnail into its soft underside. When I pulled her towards me with all my strength, she backed a few paces, thus giving me a chance to escape.

Dora

The teak poles of Dora's pen soon began to rot, and Dora was quick to discover that it was only a small matter for her to tear the fence asunder. The result was that for a long time she roamed freely in the small wood behind the Zoo. Here she developed her own little territory which she never left. The boundaries of her domain were very sharp. To the north there was the barbed wire fence towards the Zoo; to the east was some dense shrub; the southern boundary was a small stream, and towards the west was a nine inch wide gutter. The total area was about five or six acres. When in a playful mood, she made mock attacks as soon as I entered her territory. All I needed to do to avoid her was to step over the gutter. Not even her trunk crossed the territorial boundary.

Near the eastern boundary of Dora's territory, where there was a spring, she dug herself a small, muddy pond in which she had her daily bath. In another place, where the blue clay came close to the surface, she dug a mud-bath. The routine was always the same. Having enjoyed herself for some time splashing in the water, she went across to the clay to have a good roll in it. Mud and dust are important cosmetics used by elephants, keeping their skin healthy, just as many humans use powder, ointments and oil.

2.

One day just before Christmas, a Landrover pulled into our drive, and some rather scruffy men came up to the kitchen door asking for me. They had a newly born elephant in the back of the car. They wanted to know if I would have it. I looked into the car. There I saw the smallest elephant I had ever seen. I got angry when I saw it because elephants are protected in Nigeria. This meant that, as President of the Nigerian Government's Wildlife Preservation Committee, I would have to call the police. I was just on my way to the telephone when one of the men managed to explain that they came from the Federal Forestry Section near Benin, and that they had committed no offence against the law. The youngster's mother had attacked a hunter and he had shot her in self-defence. They had found the infant close to its dead mother the following day. All I needed to do was to 'phone the game warden, Christian Brandt.

Contrary to what is customary among their relatives, the large East African savannah elephants, the females of the smaller West African forest elephants always leave their herd to give birth. Such lone elephant mothers are extremely dangerous, and may attack without provocation. In the present case it was quite obvious what had happened and the hunter was not prosecuted. The Forestry office had taken charge of the infant, but no-one knew how to feed it. For five days they had tried to give it bananas but it soon became obvious that this was not a suitable diet. In the end it had been decided to ask me to take charge of it.

Carefully I lifted the small elephant down from the Landrover. It could not have weighed much more than a hundred pounds. Its dry navel cord was still dangling from its belly. In spite of the rough and ready treatment it had been given, it appeared to be in perfect health. We installed it in the garage and parked our car behind the house. A bed of straw and leaves was made and a young tree trunk was securely fastened across the open garage door so that our new charge would not run away. Having done this, we began to wonder how to feed it.

Ever since it had lost its mother five days earlier, the elephant had virtually starved. It had been unable to digest the bananas it had been given. The previous experience had been that small elephant infants always died a few days after losing their mother. In other words, they had been wrongly handled. We had no idea what had been wrong. All we knew was that it was no good repeating previous mistakes. The first thing which came to mind was that the composition of the milk had been wrong. A colleague from the medical faculty seemed to remember that he had read in an article about elephants' milk that the content of butterfat was very high. He found the paper after some searching. It gave the content as 18%. A very high percentage indeed. If the amount of butterfat really was that high, we had good reason to believe that the cause of death of all elephant orphans had been starvation. Cows' milk contained only about 6%. We had no fresh milk at our disposal, only powdered milk. This we now whisked up with melted butter and water. There could be no doubt that the elephant was thirsty. It eagerly drank this mixture when we offered it from a wine bottle with a rubber tube for a nipple.

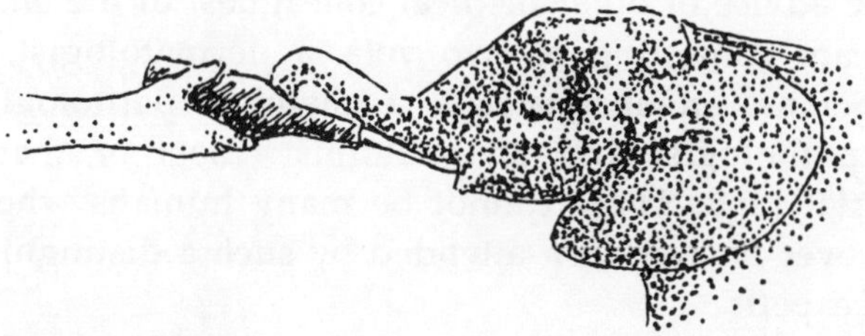

The elephant's first night was very disturbed. Many more equally restless nights were to follow. As soon as it found that it was alone, it began to trumpet. It yelled with such an energy that I feared it might disturb our neighbours. There was nothing I could do except spend part of the night on the hay pile next to it. During the first months, I probably slept just as much on the hay with our charge as in my own bed. It became a routine.

Our first experiences with the milk were not promising. It seemed to pass through the animal without being digested. From our experience with antelopes, we knew that this might cause a severe catarrh, which might become fatal. It was lucky that we had also learned how to slow down the antelopes' bowel movements by adding kaolin to the food. As it now turned out, it also worked with the elephant. By keeping an eye on its stools and adjusting the amount of kaolin we succeeded in avoiding catarrh.

After a couple of strenuous weeks, it became clear that the elephant would not die of starvation. But then new troubles were brewing. The infant began to develop severe ulcers behind the ears. During the following weeks they spread to the whole body and to the mouth. As they became worse, a yellow butterlike substance began to ooze from them. The vets could offer neither explanation nor advice. We consulted the dermatologist from the Medical School. He tried various ointments. The skin of the patient was divided into a number of experimental fields, each of which was treated differently from the neighbouring ones. Nothing seemed to help. We asked the advice of other medical colleagues. In the end we had a team consisting of two vets, a dermatologist, two paediatricians, an obstetrician, a chemical pathologist, a clinical pathologist and an anatomist, most of them of professorial rank. There cannot be many humans who can boast of ever having been attended by such a distinguished crowd of experts.

But the problem remained. Nobody could come up with any suggestions. Meanwhile I had come to suspect that it might be the additional butter that caused the ailment. I therefore decided to try to leave it out of the diet. It seemed to work, and soon after, the animal recovered. Later on, I was told that bovine fat is poisonous to elephants, and most analyses of elephants' milk have shown a fat content similar to that of cows' milk. So our information had been inaccurate, but luckily, we found out before it was too late.

Now, many years later, I have asked myself; why was I able to hand-rear an elephant from birth when everybody else had failed? There may be many explanations. Perhaps it was because it came to me so early that it was able to adapt its digestion to the strange milk. Perhaps the butter after all helped it to a good start. Maybe it was a combination of these and other factors which saved it.

One factor rarely considered is the psychological one. We know that many animal infants cannot thrive, and may die if they are denied contact with a loving mother. Even human infants will die if they are not given love. Good food, cleanliness and warmth cannot save them if they are not also given plenty of love. An elephant infant will always be with its mother. It stands and walks under her. It sleeps by her side. She caresses it with her trunk, helps it and protects it. She is always at hand and from her it gets comfort and security.

Most people attempting to hand-rear animals from infancy provide food and cleanliness in a fixed routine. During the intervening period, the animals are left alone and unattended.

The small elephant we hand-reared had its keeper and my wife during the day. I put her to bed myself at the end of the day, sat with her until she dozed off, and I often slept at her side during the night. She always had someone nearby who could console her when unhappy and to provide security. Many a time since then I have wished for another opportunity to hand-rear more infant elephants, but this is a demanding job which claims one's attention day and night. It takes an elephant about eighteen months before it can be weaned.

Although it needs less care as time goes by, the work is always very demanding. Moreover, the cost of raising an elephant is high, and it is not every day that an infant elephant, so to say, falls into one's lap. Our little elephant drank up to 65 pints of milk a day, and when it was weaned at an age of eighteen months, it had cost us £2600 (1965). Luckily, I did not need to pay all that myself. Christian Brandt was marvellous; he persuaded the Government to pay my expenses.

Because the newspapers had complained when our first elephant was given a European name, I asked a Nigerian colleague for help.

"Call it Bodunrin," he said. "It means the one that comes to the feast – your elephant came just in time to spend Christmas with you."

Thus the elephant was named Bodunrin, but usually we cut it short and just called it Bo-bo.

'Bo-Bo' aged two weeks

It is difficult to imagine a more troublesome house pet than an elephant. Elephants are social animals and, as mentioned, the young ones crave much contact with their mothers. Bo-bo, certainly, was no exception to the rule. She demanded our constant company, and it was not always an easy matter to keep her out of the kitchen. She liked to go under the table and then give it a good push from below so that everything standing on it tumbled over and fell to the floor. Out in the garden she had wrestling matches with the dustbin and the newly-planted trees. When she got tired, she often flung herself on top of an ornamental bush. She was active and inquisitive in the extreme. Everything had to be investigated and tested. Her small trunk was never at rest, and seemed to get in everywhere. Nothing left on the window-sill was safe. Just like a snake, it silently crept in through the window and made its way in between the various ornaments. It played with them and twisted itself around them. One by one, they disappeared out into the garden.

At one stage, Ibadan had a severe water shortage, and we were all instructed not to waste it. Having Bo-bo around, we found it impossible to adhere to the restrictions. The garage, where she slept, had to be washed and scrubbed daily, and Bo-bo herself had to be washed and scrubbed. She loved water and never tired of playing with it.

To keep Bo-bo occupied and to satisfy her preoccupation with water, we bought her a small plastic bath. In the beginning she would spend hours sucking up water into her trunk, seemingly with the intention of pouring it over her back. It did not work too well and one could almost see her disappointment every time she lifted her trunk out of the bath and the water ran out before she herself could blow it out where she wanted it to go. Apparently she had to learn the art of retaining it. When she finally accomplished this, she added new elements into her games. For a long time she experimented with the garden hose, picking it up, trying to drink from it, and at the same time to suck water up into her trunk, with the apparent intention of pouring it over her back. A very complicated action indeed.

As long as the bath lasted, it gave her endless amusement and we enjoyed seeing her play in it. She climbed in and out of it, splashing to her heart's content, but unfortunately there is a limit to how much rough treatment even a strong plastic bath can stand up to. In the end it cracked and that put an end to the game.

Meanwhile, Bo-bo had discovered how to turn on the tap, and that became her favourite game. Of course, she never bothered to turn it off again. Just as quickly as we stopped the flow, she was back again, wasting more of the precious commodity.

The desire to play with water seems a characteristic of elephants. The former Director of what in those days was known as the Murchinsons National Park in Uganda, "Bombo" Trimmer, told me many years ago about the problems he had with the then famous elephant, Lord Mayor.

Lord Mayor preferred to spend his time in the vicinity of the hotel. The hotel manager, Toni Nuti, a lady who could not have cared less about laws and regulations, frequently fed the elephant with refuse from the kitchen, thus encouraging it to hang around the place. Never mind how much Trimmer pleaded with her, Toni would not listen.

Of course, Lord Mayor discovered how to turn on the water at the taps. Water was expensive, and, after some time, Trimmer decided to put a lock on all the taps. Next morning when he came out, the camping site was flooded. Lord Mayor had amused himself tearing up the water pipes and breaking them. It was then suggested that it would be necessary to shoot Lord Mayor, but Trimmer resisted. He was too fond of this magnificent animal. Some time passed, until one day Lord Mayor came to Trimmer's house and stole a bottle of whisky through the open window. This settled it! Trimmer grabbed his gun and followed the thief out into the bush. Thus ended the life of the most magnificent elephant they ever had in the park. Mrs Trimmer added to the story by telling how her husband came in with tears running down his cheeks.

The attitude of "Bombo" Trimmer towards Lord Mayor seems to be typical of those who work with wild elephants. Once I asked a game ranger who had been a professional hunter whether he was not afraid when shooting elephants.

'I am dead scared,' he admitted, 'but I love them all the same. I can't bear to kill them.'

His love of elephants led him to give up hunting them. His job was now to protect them.

To the Nigerians, Bo-bo became a great attraction. We had streams of visitors who wanted to see her. Obas, chiefs and cabinet ministers called on us. Even the King of Ondo and his Queen came for coffee one day to see the elephant. They came escorted by police on motor cycles with screaming sirens. Such visits drew even more attention than the elephant they had come to see.

When Bo-bo was nearly seven months old, and I had to go on leave to Europe, we decided to transfer her to the Zoo. She had by now grown so big that it was difficult to keep her at home. Much of the garden was in ruins and the garage door was splintered. Early one morning, before there was much traffic on the roads, we walked her over to the Zoo and gave her a pen of her own, close to Dora. She mourned deeply over the loss of her first home, and it was necessary to tell the keeper who had helped with feeding and looking after her, to spend most of the day in her pen. It was touching to see her follow him wherever he went. Usually he carried a stick under his arm. Bo-bo held on to this with her trunk in the same manner in which wild elephant children hold on to their mother's tail.

The thought of introducing Bo-bo to Dora had filled us with anxiety. Nobody could tell in advance how she would be received. As it turned out, our fears were unfounded. We arrived one morning to find that during the night Dora had torn down the fence between the pens. The solid teak poles and the heavy concrete blocks in which they were fastened were strewn all over the place.

They had themselves solved all the problems and stood peacefully side by side. From then on they remained good playmates.

When, a year later, we had a new pen for them on the other side of the brook, we moved them over there. It was a good strong pen, built from railway rails welded together. The move was, however, more difficult than we had anticipated. The animals did not want to leave their established territory. In the end we had to throw Dora on to a pile of hay. It took seventeen strong men to drag her across the small bridge into her new pen. Both elephants settled down almost at once in their new surroundings and from then on we could let people see them at a closer range. They became one of the zoo's major attractions.

Dora's character was very different from Bo-bo's. Dora was gentle and easy to handle; Bo-bo was disobedient and sometimes malicious. When elephants show affection for one another, they often put their trunk into one another's mouth to touch the palate. It happened, on some occasions, when Dora showed her affection this way, that Bo-bo bit her. Once she also bit my elder daughter. It was sheer luck that she did not lose her finger. An elephant's back teeth are like mill stones. The girl only lost her nail, and this was painful enough.

I suspect that the cause of Bo-bo's bad behaviour was that I had spoiled her. An elephant mother is not above giving her child a good cuff if it misbehaves. Bo-bo was five months old when, for the first time, I lost my temper with her and dealt her a good blow with my fist on her forehead. Of course, this was without any effect whatsoever, but I got a sprained wrist. From that day on I preferred to use a stick, but as soon as she saw it in my hand, she fled, trumpeting, and still refusing to obey orders.

Luckily I had some excellent keepers for whom the elephants showed the greatest respect. I remember one day some years later when I wanted to give an avocado to Bo-bo. A recently-employed keeper was with the elephant. He stood in the far end of the pen, with an animal either side. I called Bo-bo. Contrary to what usually happened, she did not stir. I called again. Still she did not move. When I called for the third time she suddenly swung round and hurried over to me.

Later that day I asked the keeper why Bo-bo had not come when I called her.

'I did not want her to until I had told her that she could,' came the reply.

I never got the same control over the elephants myself.

It is said that elephants never forget. When I returned to Nigeria after having spent four years in America, I was thrilled by the reception I was given by several animals who apparently remembered and recognised me.

Both elephants put their trunks around me and gave me a big hug. They would not let me go until I had patted them and rubbed their palates.

There were other animals too who seemed to remember me. A duiker antelope came up to the fence as if to greet me, and an old patas monkey called me with her soft voice. She recognised me from a great distance. We had been friends before I left and now it was as if we had never been parted. She and I remained friends. She was a peaceful, harmless animal towards me. It became a daily routine that I visited her and we groomed one another. It was not until shortly before I left Nigeria for good that a keeper informed me that she was a very dangerous animal.

BABOONS AND OTHER MONKEYS

Although my Doctor's dissertation was on entomology, I ended up studying monkeys. This change in my career came as a surprise, even to myself.

My research into insect physiology had brought me on to problems which demanded the use of very delicate and expensive equipment. I was able to invent and construct the simpler instruments myself, but there were many other things which I could not make, and I had no money to buy them. Some colleagues who had better facilities than I offered me some help but, as it meant some effort on their part, it never materialised. My eyes also caused me some trouble. I felt thoroughly frustrated.

One day I was sitting in my office when the telephone rang. Professor Raymond Dart wanted me. He asked me to come to his office in the medical school. I felt the greatest admiration for Dart and, for some inexplicable reason, he was showing some interest in me. Although Dart was Professor of Anatomy, and was especially interested in neuro-anatomy, circumstances had brought him into physical anthropology and certain aspects of cultural anthropology. In the early twenties he had described and named the first find of Australopithecus. Dart insisted that this creature, which lived more than two million years ago, was a primitive human. This view was strongly opposed by all the well known anthropologists at the time. They thought it could be nothing else but a new and interesting ape. It was not until many years later, when most of his opponents had died, the second World War was over, and Robert Broom had found another almost complete skull that the world had begun to come round to Dart's view.

I admired Dart for the tenacity with which he fought for his convictions. He was unusually imaginative, and the spirited battles he often fought with those in authority had benefitted many of us younger university employees.

When I arrived in Dart's office, he told me that the day before, he had had a discussion with the well-known expert on the Kalahari Bushmen, Marshall, of the Smithsonian Institute in Washington. There were many aspects of the bushmen's behaviour they could not explain. They had finally agreed that a study of a monkey species, such as, for example, the baboons, might throw some light on the problem.

What Dart wanted to know was: could I look into the possibility for such a study, and give an estimate of its cost? Secondly, would I consider carrying it out myself, or at least supervise it? Money would be no obstacle, and he expected the study to run for several years.

This proposal astounded me. How could I, an entomologist with no idea about monkeys and their behaviour, undertake such a task? He even asserted that there was no-one else in the whole of South Africa capable of conducting a study of that kind. A few years later, I learned that there was someone else in South Africa who was better qualified, had more time at his disposal and had more money than I to conduct such a study. It was Ronald Hall, who at that time was Professor of Psychology in Cape Town. Later on he became Professor in Bristol where, most unfortunately, he died after being bitten by one of his monkeys.

Dart's ability to persuade was remarkable, but he was a man prepared to take a few chances, perhaps even to be a trifle reckless. When I left, he had succeeded in making me promise to consider his proposals.

It was just before Easter, and my family and I had planned a camping trip to Natal.

'All right,' I said, 'we will change our plans. Where do you think we can find a good place to study baboons?'

Dart suggested Waterberg, in the northern Transvaal, where Eugene Marais had observed baboons before. He knew a farmer there.

Two days later, my family and I were on our way to Rankins Pass in Waterberg, and with that I entered a new epoch in my career.

The time at Rankins Pass was a new experience for us all. There were plenty of baboons; several troops of them. They gave me a good idea of their daily routine and various aspects of their behaviour.

I spent a night on a shelf on the rocks about 200 feet from a deep cave which was the hide-out of a leopard. Very little happened that night, apart from a thunderstorm with gale-force winds that nearly blew me and my tent down into the valley.

Back in Johannesburg, I reported on my experiences to Dart and Marshall. The latter I found sitting on his hotel bed at eight o'clock in the morning, still in his pyjamas, sipping his coffee. The estimate I gave him was for a day-to-day study, stretching over a period of seven years. The estimate included the purchase of two non-profitable farms and the cost of various electronic equipment, salaries, transport etc.

Marshall listened with great interest, and called his son to take part in the discussion. I was assured that money would be no obstacle and I left full of great expectations. Why Marshall never acted, I cannot tell. Two years later, when I met him again, he was still talking about finding someone to study baboons. He seemed to have forgotten all about our conversation that morning in his hotel room. I had already decided to do things without his help. By that time, I was deeply engrossed in a study which I had financed myself. But his plans for a primate research centre never matured.

Since then I have stressed the importance of a centre for the study of wild baboons. Once, in another African country, there were plans that an area of nearly one hundred square miles of wild bush land, with the necessary buildings, should be set aside for the purpose, but the plans collapsed for political reasons.

2.

Shortly after Dart suggested that I should take up the study of baboons, I had a large strong cage erected in our garden, and the Institute for Polio Research kindly presented me with two small animals. My reason for wanting to have baboons at home was that I did not want to venture into a great expensive project without knowing how to tackle the animals. The better prepared I was, the less time I would waste.

The two baboons, which we named Joe and Jenny, were very affectionate and gave us much joy. During my daily acquaintance with them I was able to acquire a reasonably good insight into their system of communication in a very short time. Later on, this proved invaluable. However, it became clear to me that a cage like the one I had for my animals offered only very limited possibilities for useful credible studies of behaviour. I felt very frustrated, and was more than grateful to Labuschagne, of the National Parks Board, when he arranged for me to have special facilities in the Kruger National Park. This made it possible for me to carry out field research. The observations I made were of the greatest value to me in the coming years. A couple of minor publications I wrote were the first to be published on the ecology and behaviour of wild monkeys south of the Sahara. These publications were, shortly after, to be followed by some more comprehensive papers by Ronald Hall.

3.

It has always been my conviction that it is by making comparisons that one best becomes aware of the problems. Having had Joe and Jenny for some time without having much to show for the experience, I therefore decided that it might be useful to acquaint myself with the behaviour of other primate species. It was lucky that Dr Paulian from the Scientific Institute of Madagascar in Tananarive came to visit us. I told him of my problems and he promised to let me have a pair of ringtailed lemurs. These animals were relatively common in the sourthern part of the island, where the climate is similar to that of South Africa.

It was not long before a couple of beautiful animals arrived. They too were given a large cage in the garden. After some time, when the animals had become well acquainted with their new surroundings, I allowed them to roam freely in the garden during the day, where they hunted for insects. They were fed in the cage and returned to it every night to sleep. Keeping the lemurs in a semi-wild state was a successful experience. In Madagascar they live in open bush land, and because we lived in a dense eucalyptus wood, they never ventured too far from our garden. Their territory extended from the bottom of the garden to the middle of the road in front of the house. They bred well, having a single infant or twins every year, and soon we had a nice little troop.

Apart from being social, the lemurs turned out to have very little in common with the baboon. Their mode of communication was largely by means of odours. They were endowed with a number of scent glands, situated on shoulders, arms and genitalia. These they made use of when threatening, warning, marking territory or making themselves known to each other or the opposite sex. Their social organisation differed from that of the baboons. Outside the breeding period, the females seemed dominant over the males.

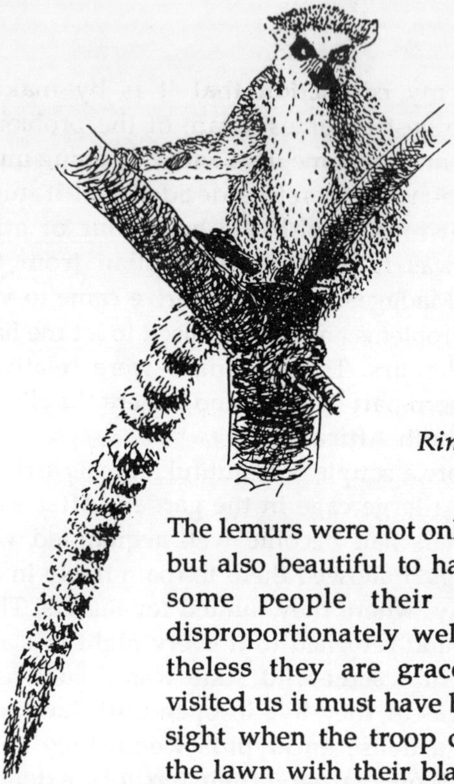

Ring-tailed Lemur

The lemurs were not only interesting animals, but also beautiful to have in the garden. To some people their hindquarters seem disproportionately well developed. Nevertheless they are graceful. To many who visited us it must have been an unforgettable sight when the troop came strolling across the lawn with their black and white ringed tails curved over their back in an elegant "S".

Unfortunately, the lemurs could be dangeous to have around, especially during their breeding season, and on several occasions we were attacked. Luckily no severe harm was ever done.

Later on I was invited by Dr Paulian to study lemurs in Madagascar. It had been my intention to go to the South to look at the ringtailed lemurs but weather conditions forced me to change my plans. Instead, I moved North to the island of Nossi Be, where I acquainted myself with the forest-living black lemur. The males of these animals were as black as black cats and had yellow eyes. The females were as red as the European fox and had white tufts on their ears. The females seemed dominant over the males and were more aggressive. I was constantly threatened by them. At times they sat as close to me as six feet, uttering their warning doc-doc, while swinging their long tails like pendulums.

The adaptation of the lemurs to life in the trees was to me astounding. Effortlessly they ran up the tree trunks and along the branches, and they were practically never seen on the ground. When moving from tree to tree they often made colossal jumps. Sometimes such jumps could be more than twelve feet; indeed a remarkable effort of an animal smaller than a cat. My ringtailed lemurs at home did not seem capable of managing quite such long jumps.

I had my ringtails for many years, until the whole colony was wiped out by an epidemic. When we decided to leave South Africa and move to Uganda, I took the lemurs with me. This transfer from the southern hemisphere to near Kampala in Uganda, about thirty miles north of the Equator, had a surprising effect on them. They changed their breeding time to fit in with the northern hemisphere. I mentioned this to colleagues in America and elsewhere. The response came as rude letters. Nobody would believe that a move to a locality thirty miles the other side of the Equator could possibly be enough to induce changes in the breeding cycle. But then an American ornithologist told me that he had noticed that some bird species living along the Equator had breeding cycles which varied between those living north and south of the Equator, even when the distance to the "Line" was only a matter of thirty miles. Those north of the Equator had the northern, and those south of the Equator the southern breeding cycle. I felt vindicated. A year later, I had a letter from America. The writer apologised for his previous rude remarks. He had now made the same observations. I am still unable to understand how it is possible that the small differences between the length of days in two places separated by only a few miles could have so profound an effect as to change the breeding cycles.

Vervet

4.

Now I had begun to make comparative studies I decided it would be useful if I also acquired a pair of vervet monkeys. For those, I built a flimsy cage close to the baboons. The thin chicken wire I used for their cage was not nearly strong enough and the two active animals continually escaped and spent days on end in the top of the trees. It was only when they were hungry or they saw something to steal that they ventured down into the garden. Although I enjoyed seeing the animals move about freely, my pleasure was not always shared by my wife. If someone forgot to close a window, or the door to the kitchen, nothing was safe. They never missed the opportunity to grab food that was left unguarded.

At that time our domestic help had her boyfriend Nathaniel living with her. Thanks to my wife's initiative, he had recently had his mouth cleared, in the dental school, of a set of decaying teeth which caused him much pain and discomfort. He was given a whole new set of false teeth, in which he took much pride.

Whether it was to save his new teeth from wear or for some other reason I don't know of, Nathaniel often took them out and placed them on top of a high wall connecting the kitchen with the servant's quarters. As could be expected, they were irresistible objects to the vervets, and one day they saw the opportunity to steal them before anybody could interfere. They quickly climbed to the top of the roof, where they settled on the ridge, stuffing the teeth into their mouths.

It looked comic, but the loss of teeth spelt disaster for Nathaniel. Without hesitation, he and a friend pursued the monkeys on to the roof. Of course the thieves disappeared down the other side of the house. A wild chase ensued. Nathaniel and his friend ran forward and backward in the flower beds and the kitchen garden, trampling down flowers and vegetables. In the end the monkeys dropped the teeth. Nathaniel retrieved them and put them, unharmed, where they belonged.

It was not only the vervets that sometimes escaped; sometimes the padlock of the baboon cage failed to click. On such occasions Joe and Jenny were sure to discover it, and then took a day off in the garden. If the two animals had not been so destructive we might have allowed them free range, but they seemed to enjoy ruining whatever came within their reach. It was wonderful to see how happy they were galloping all over the garden. What was more, our big dog did not even protest when they took a ride on his back.

The baboons' escapades would have been less alarming if they had confined themselves to roaming in the garden. Apparently they found it more entertaining in the house. We could be certain that hell would literally be let loose if we did not manage to close all doors and windows before they got in. There was no limit to how much they were capable of ruining in next to no time. They were well aware that we did not approve of their behaviour, but that only served to entice them to further misbehaviour. The situation invariably deteriorated when we tried to capture them – and capture them we had to. As soon as they noticed it, they got nervous diarrhoea. There is no reason here to repeat what my wife had to say one day when the baboons got into my daughter's room. They were enjoying themselves by swinging on the curtains and falling down on the soft bed.

When they saw us entering the room they panicked, and before we had them under control, they managed to smear curtains, walls, bed, furniture, books and floor with excreta. It took us hours to restore order, and the smell lingered for days.

5.

Monkeys are definitely not suitable as pets. All monkeys bite, and monkey bites are dangerous and may be deadly. Monkeys are not dirty by nature. Moving in the tree-tops they need not worry where their droppings fall. They do not build nests which must be kept clean, as birds do. Consequently, cleanliness with their stools has not become part of their natural behaviour, and they cannot be house trained.

Wild monkeys live in troops; individuals are organised in a hierarchy, often established by fighting. There is competition and each animal has to assert itself and defend its social rank. In the same way as among humans, the individual's standing may rise if it is able to do something which no-one else can or dares attempt. To do something outrageous or non-permissible, and get away with it, often inspires respect, and adds to an individual's authority and status, as do possessions. In the world of monkeys, status and authority are very important. Monkeys will obey the one who wields authority, as long as he is present. In his absence, they will do as they please. It is these characteristics which make it impossible to have a monkey in the house. No-one can be certain of what they will do next.

6.

Shortly after we had moved to Uganda, a local man called to ask if I would like to buy his tame female baboon. He had to part with it because it was addicted to stealing small children. Although the animal caused no harm, no-one could blame the man or his wife for feeling anxious when the baboon sat on top of the roof, cuddling their baby. I bought Maco – that was the name they had given the animal – for five pounds. This was the most I had ever paid for a monkey, but I soon found that this one was well worth the price. Maco revealed some of her peculiarities as soon as I took her to my car. I offered her a banana, but instead of eating it she sent it flying, with a remarkable precision, at a man she apparently did not like. Towards me she was very polite, but at home she tried to establish her authority over my elder daughter, Janet, by giving her a bite on her fingers. That, of course, we could not accept, because it was beyond any doubt that Maco could be dangerous if not kept in check. I explained this to Janet, and told her to give the animal a good beating should it again try to assert itself.

Of course Janet had another bite. She did as I had told her, but, being of a gentle nature, she did not have the heart to hit Maco hard. I scolded the girl and told her that baboons were tougher than she thought, and that she need have no fear that she would break any bones.

Next time Janet was bitten, the skin on her fingers was broken. I was close by ready to assist her. This time Maco had the beating of her life. Janet hammered her with all her strength, while I held her firmly. From that day on, the two were great friends, and Maco never made another attempt to bite Janet. Their friendship lasted for many years until I had to part with Maco. When Janet later on went to boarding school in England, only coming home to Africa at Christmas, Maco was always the first old friend she had to visit. It was a joyful reunion every year, with hugs and kisses, and their friendship was as close as ever. No-one else but Janet could handle the, by now, very dangerous animal with the same confidence.

Having seen how skilful Maco was at throwing, I began to test her abilities. Not only was she able to throw objects with great precision, but she could also rake in tit-bits with a stick. She liked experimenting with whatever she got hold of, and while playing with sticks, she discovered that by balancing them on their end, she could climb them and reach food above her which was otherwise out of reach. Her ability to analyse and comprehend a situation and solve problems was phenomenal and seemed hardly to lag behind that of chimpanzees. In the world of baboons she was a genius!

I felt sorry that we never succeeded in breeding from Maco. She was very maternal and loved to have something to cuddle.

Maco with a young serval

She never missed an opportunity to kidnap an infant animal, and we had many anxious moments when she sat in the top of her tree with a squalling kitten or a valuable puppy. Although she never caused them any harm, she always made us feel uneasy. However, having grown up among humans without the company of other baboons, she had not learned the proper baboon behaviour and did not know how to socialise with animals of her own kind. All attempts to introduce her to a male failed.

One day as I sat in my study in Uganda, I suddenly became aware of numerous small bits of paper fluttering in the wind past the window. They all seemed to come from the direction of the tree where Maco was tethered.

I went out into the garden to see what was going on. It was as I suspected. Maco sat in her tree with a bunch of papers which she tore into small bits. I picked up some pieces. They were parts of grocers' bills, which their van driver was supposed to deliver to the customers. It was to no avail to scold Maco, who apparently was delighted with her new toy.

Only the driver of the van was to blame. He had parked the vehicle where he knew he was not supposed to park, and had left the bills on the seat, in full view through the open window, while he delivered the wares. It was too much to expect that Maco should be able to resist stealing anything so obvious. Luckily, I managed to retrieve all the fragments. The grocer punished the driver by refusing him his wages until he had pieced all the many bits together. It took him a whole month to do it.

Not surprisingly, our neighbours disliked Maco and some were just waiting for the day to come when something happened which gave them grounds for demanding that Maco was removed from our garden. One day it really looked as if their wish would come true. Two men called at our neighbour's and asked for help. They had many deep wounds, were bleeding profusily, and one had lost part of an ear. They claimed that they had been attacked by Maco. When the neighbour's servant heard the story, he burst out laughing, something for which he was severely reprimanded. This was a very serious matter. Maco would have to be removed without delay. Despite our neighbour's anger, her servant could not suppress his mirth. Infuriated, she demanded an explanation. Why was he laughing? "Memsahib" came the polite answer, "it was not Maco who bit them. They have been fighting and they bit one another." The two men had been drinking and started a fight, during which they gave one another some good bites. But they were ashamed and decided to blame Maco.

It was beyond doubt that Maco could be very dangerous. She was unusually intelligent and mischievous, but her sense of humour was often crude. When angry, she would throw stones at the offender. Her aim was precise and it was our luck that no-one was seriously injured. Most of the time she was friendly and loving, and we all got very fond of her. She had more character than any other monkey I have known.

7.

On a trip to Serere, near Soroti in Uganda, I was presented with a tiny male infant Patas monkey. He must have been only five days old. We named him Freddy. Hoping to gain a deep insight into the many small facets of the monkey's behaviour, I decided to try myself as a monkey mother. Thus, Freddy's and my life became closely integrated for a period of many months.

Our first experience together had, however, very little to do with monkey biology. We were only about half an hour's drive from Serere, when a boy carrying a bucket of water on his bicycle lost his balance and fell in front of our car. There was nothing I could do except drive into the ditch to avoid running over him. Our lovely new camping car heaved over on its side and ended with all four wheels in the air. Nothing happened to the boy. He took fright and fled into the bush, leaving bicycle and bucket in the road. But I had been injured. A spanner, which normally stayed under the passenger's seat, had knocked a hole in my head. Blood was streaming down, worse still, it was getting mixed with hot oil from the engine as it ran down over my face.

'Turn off the engine,' commanded my wife. How could I? She was sitting on top of me and I could not move. But I managed to move a little with some difficulty. She shifted her position so that I could get a hand free and switch the engine off.

'Where is Freddy,' asked my wife.

'I have no idea.'

At long last my wife's brain began to function a little more normally and she suddenly remembered that our younger daughter was in the back of the car.

'What's happened to Kirsten?' she enquired.

'I am all right,' came a faint voice, 'and Freddy's all right too.' Kirsten was buried under all our bedding, furniture, cooking utensils and camping equipment.

Feeding Freddy

After much effort we managed to get a door open and get out. A group of villagers appeared from nowhere and kindly helped us, getting the car on to its wheels and back on the road. It was a sad sight. The car, which had been our pride, was in a terrible mess. Some of the nice furniture was splintered. But we could still drive it, and we were all well, apart from the hole in my head, which was not serious.

Some time later, when I was on my way home with another new monkey I had just collected, the car skidded in mud on its way up a hill and turned over a second time.

It stands to reason that I was not a good monkey mother. As Harlow pointed out shortly afterwards: maternal behaviour must be learned. Being of the wrong sex, I lacked the innate foundation for the acquisition of such behaviour, and I spoiled the baby with my anxiety. Wherever I went, Freddy was always with me, and he became a pest to many people. I had him with me when lecturing and at parties. He even went with me to the cinema, and sat nicely through "The Ten Commandments". I smuggled him into the theatre hidden inside my trouser leg. At night I tried to make him sleep in a cardboard box next to my bed, but that he did not like. He wanted to sleep in the bed next to me, so in the end I gave in. Thereafter his place was on my pillow. He always slept with his arms round my neck and the lobe of my ear in his mouth. He rarely dirtied the bed. As soon as I woke up in the morning, I took him with me to the bathroom and held him over the loo, just as we do to human children in a similar situation. It worked splendidly.

During the first few months I encountered very few difficulties when I went shopping with Freddy on my shoulder. He always held on to my hair or to my ears and stayed there. But from the age of four months he became increasingly adventurous and began to leave me to explore while I went on with my business. At first these small excursions caused no problems. It was not until one day I had to call at my bank that things began to go wrong. If Freddy would have stayed on the customer's side of the counter all would have been well, unfortunately he became fascinated by all the things the teller had in front of her and went over to her side to play with them. The teller did not seem to mind and appeared to find him amusing. After a few moments, however, Freddy saw the opportunity to steal her pencil and would not give it back. He jumped down onto the floor and ran into the nearest office taking no notice of my calls. A major chase ensued during which Freddy panicked and ran from office to office causing havoc by knocking papers onto the floor. Finally he escaped through the front door. Out in the street he caused a minor traffic jam before he finally spotted me and jumped onto my shoulder for protection.

As time passed, Freddy became increasingly troublesome. Everything had to be examined by him: touched, smelled, stuffed into his mouth and tested in numerous ways. He learned the characteristics of the object, but it also meant that nothing in the house was safe against his explorations. While I was lecturing, he would swing gaily on the lamps, and then in an unexpected jump, land on a table stealing the student's pencil or notebook. Soon he began roaming about. Troupes of patas monkeys always tend to spread out, as they live in the open savannah where they can easily keep an eye on one another. When Freddy found life dull, he often visited the geology department, where the many different stones seemed to offer a special attraction. He could have long sessions sitting quietly playing with them. Luckily, the Head of Department, Dr Henderson, found it amusing.

Wherever Freddy was, he was always carefully watching my movements. If at any time I got out of his sight he became very anxious. One day when I had a chore to do in the workshop, he refused to come when I called, so I left without him. A moment later I saw him searching for me on the lawn. He looked panic stricken, and I could hear his soft frightened cry. Having watched him for a few moments, I called him. As soon as he heard my voice and knew where I was, he galloped across the lawn and jumped in through the open window. The next moment I had his small arms round my neck and my ear had a mighty big suck before he dared venture out on other adventures.

It has a calming effect on a small monkey when its mouth is brought into contact with something soft. In this respect monkeys resemble humans. My earlobe gave Freddy comfort, as the dummy does to the human child. It was quite common at home that Freddy suddenly broke off a wild game with the other animals to come and embrace me and give what virtually amounted to a kiss on my mouth. It was all very brief; he had to be off again to play. It was such small signs of affection which I found difficult to resist.

To him I was a trusted object, available for soothing contact whenever Freddy became frightened or over-excited.

It is understandable that my family did not entirely share my love for Freddy, and he was often the cause of some very rude remarks from my wife. He would not leave her things in peace and stole whatever he could lay his hands on. He was particularly addicted to attacking her flower decorations, parts of which he liked to eat. This, of course, we could not allow because it invariably resulted in the bowl landing on the floor. On the other hand, we found it useless to punish him because this would only emphasise the value we placed on the object he wanted to have. The more we tried to prevent him from touching something, the greater became his desire to possess it. We therefore attempted to distract his attention by pretending that some other object was of an even greater value. This could be difficult to achieve on the spur of the moment, and there were a fair number of breakages. Yet we succeeded now and again, and I shall never forget the day I distracted him from pulling down a vase by drawing his attention to a "very valuable" roll of toilet paper.

From time to time I had to travel to other parts of Uganda or to Kenya. On such journeys I was always accompanied by Freddy. To him it seemed a great experience. There were new impressions and many new things for him to explore. As long as we moved in open country he always seemed quite at ease, but when we came into deep forest, where the close vegetation hindered a free outlook and the sky was obscured by the canopy of the trees, he appeared frightened and ill at ease. He would then keep close by my side or ride on my shoulder.

8.

At Christmas we decided to visit Toni Nuti, the same lady who contributed to the Lord Mayor's death.

Toni lived on an island in the middle of the cataracts of the Kagera River, the river which forms the boundary between Uganda and Tanzania. The access to her island was along a tree trunk which spanned a minor waterfall. We found her house to be in a beautiful spot, peacefully tucked away from all daily turmoil. Being in a remote and inaccessible place, Toni had succeeded in developing her islands – she had several islands, all connected with trunks of felled trees – into a small private nature reserve. There were monkeys in the trees and otters in the river. There was also a small herd of hippos. The old hippo bull, Napoleon, was the unchallenged and awe-inspiring master of his domain. After dark, it was advisable to remain indoors, or at least not to stray too far away from the house, so as not to stir the anger of His Majesty.

Toni Nuti was a very exciting personality. She had left her home in Italy and emigrated to East Africa when she was young. For some time she earned her living by washing gold from the sand in the swamps and rivers. Later on she found other ways of making a living, and for a number of years she held a job as labour-recruiting officer. I believe it was during this period that she befriended the Sultan of Zanzibar, and when we got to know her she had her house filled with valuable presents from him. When Toni got tired of recruiting labour, she succeeded in getting the job of manageress of the

Safari Hotel in the Murchinson National Park. I have no idea when she married, but she did not live with her husband. As a devout Roman Catholic she could not divorce him. She had for many years had a boyfriend, with whom she hoped, one day, to settle down. When the two of them thought that they had saved up enough capital, they decided to build a hotel on the islands in the Kagera River. The foundations were laid, and the house intended to serve as their private residence was completed, but having gone that far, they ran out of money and the hotel was never built. At the time we visited Toni, she ran her house as a small holiday guest-house.

It was New Year when we arrived at The Islands. All the other visitors she expected had sent their apologies for not going due to the political unrest which threatened to develop into civil war. Her finances were at that time in a very precarious state, so she was overjoyed when we arrived.

It was rather naughty of us to arrive as we did, without warning her that we would bring with us two servals, two pekingese dogs and Freddy, the monkey. Toni was, naturally, taken aback when she saw us balancing across the tree trunk with our menagerie. She was concerned that her animals might not give ours a good reception. We had, however, no difficulties in convincing her that she had no cause to worry, and we promised her that our animals would not leave the house after dark, in case Napoleon should be around.

Freddy was luckily on his best behaviour during our stay, and Toni Nuti was very understanding when, now and again, he transgressed. She had herself had baboons who occasionally came in through the window and upset everything. It happened several times while the visitors were having their tea.

Not all the visitors appreciated it when the baboons came and sat on their shoulders, ate their cakes and galloped over the table, spilling the tea. In the end she had to shoot them.

The little servals caused no trouble. They stayed in our room, except when we took them for a walk. Toni's twenty-year old dog was too feeble to bother about our animals, and her other dog soon accepted them as part of the household.

9.

It had been my aim to let Freddy grow up to become a real patas monkey. I did not want him too humanised. Luckily, I had acquired more patas monkeys, which I kept in a large cage in the garden. Whenever I had reason to enter their cage, Freddy came with me. This way he learned to socialise with monkeys of his own kind.

As time passed, it became obvious that I was too protective in my attitude towards him and too prepared to give him comfort when he asked for it. Many years later, I noticed that monkey mothers "ration" the comfort they give their children. One thing I noticed was that the mother often placed an arm or a hand between herself and the clinging child, so that it could not get as full a contact as it wanted. My impression, from many years of observation, is that infants become easily addicted to soft, warm contact. It seems to have an almost narcotic effect on them. Outright rejection seems always to have an effect which is the opposite of the intended. The child becomes clinging, mother-bound and retarded. But gradual weaning from warm contact has no such effect. Later on, I also noticed that the possibilities of socialisation with peers helps the young to become independent. The mothers are more permissive towards their male infants than towards their female infants, and allow them to socialise earlier.

The movements of the female infants are restrained more. Often they are held back and they are not allowed to stray as much as their male siblings. They become more mother-bound. At a very early age the monkey children begin to show interest in younger infants. They like to hold them and play with them – something the males do much less than the females. The result is that small infants get more attention from older sisters than from brothers. An infant with an elder sister learns therefore to socialise earlier and more easily than an infant with an older brother.

I noticed interesting differences in one respect between the behaviour of most human societies and that of monkeys. Among humans, the first-born child tends to hold a status superior to his younger siblings. This in particular applies if the first-born is a male.

Patas Monkey

Among macaques and baboons the opposite is the case. The younger child dominates its older siblings. The reason for this is that the young infant, which cannot fend for itself, is entirely dependent on its mother's protection, and is therefore identified with her. The older child, which remains dominated by its mother, but is no longer identified with her, is therefore also dominated by its younger sibling. To rebel against this order of dominance may result in severe retribution from the mother and from other monkeys, and for this reason it is not easily changed.

I once had a young male monkey who had learned how to wash his food before eating it. Every morning when I fed my monkeys with carrots and kohlrabi, he gathered as much as he could carry, took it to the drinking trough and dumped it into the water. Each item was then carefully washed. This same monkey had a younger brother who had not learned the trick. He just waited until big brother had done the work and then helped himself to as much cleaned food as he wanted. Big brother never protested. An alarm cry from his younger brother would have set off an attack from his mother and from the other monkeys, and that he could not risk.

The superiority of the younger child over its older siblings can, of course, have a fundamental effect on their social structure as a whole. Among the baboons and macaques, the males usually break free of all female dominance, except that of their mother. By fierce competition they organise themselves in a social system dominant over that of the females. The position of each male is determined by its strength, courage, intelligence and experience. The strongest and most cunning animal may become the leader. Among the females of most baboons and macaques the dominance order acquired at birth seems more permanent, but I have seen some indications that it may be changed, with age, through competition. The pattern of organisation mentioned here is not valid for many other monkey species. More long-term observations are required to unravel the various principles.

Monkeys are always prepared to defend someone from their troupe who is in distress – especially if it is a young one. In this respect they are similar to humans. I remember how, as a boy, in blind fury, I attacked a man who wanted to drag off my brother, whom he thought had been stealing strawberries.

One day Freddy brought his favourite kitten with him into the monkey cage. The old monkeys were very interested in the kitten and started grooming it. Suddenly the kitten gave a small mew, such as kittens do when handled a little roughly and are prevented from escaping. This triggered off a furious attack from Freddy. Wild with rage, he flew right into their faces, scratching and biting them. The older monkeys were so flabbergasted that they dropped the kitten and allowed me to pick it up and put it outside.

As Freddy got older, I began to leave him for periods of time in the cage with the other monkeys. This, however, did nothing to promote his independence of me. Although he seemed to behave normally when I wasn't there and played nicely, he always became unhappy when he saw me, cried and wanted to come out. I felt that by spoiling him and being over-protective I had created obstacles to his emancipation. Usually a monkey child has to vacate its place in its mother's lap at the age of one year, when a new sibling is born. This is always a traumatic event which causes temporary retardation. The child becomes more clinging and mother-bound. The retardation is greatest when the mother has been over-protective and over-indulgent with bodily contact. Since, however, it is only the access to its mother's lap it is denied, and other possibilities for contact remain as available as before, the child usually recovers within a few weeks. In Freddy's case, the enforced separation was complete and therefore more traumatic. When in the cage he had no access to me; I was out of his reach. However much I tried, his ties to me, the only "mother" he remembered, would not break. He preferred my company throughout his much too short life, which was terminated at an age of three years when he was killed in an accident.

I have had many monkey children since Freddy died, but I have never attempted a similar close relationship with any of them. Their upbringing, apart from their feeding, has been left in the care of other monkeys. A human, even with the best intentions, cannot replace a monkey mother.

THE APES

It was again Professor Dart who helped me to become acquainted with the big apes.

L.S.B. Leakey had, in the second half of the 1950's become interested in the mountain gorillas on the Veringo Volcanoes along the border between Uganda and Congo (Zaire and Rwanda). To gather more information about the possibility of a more systematic study of the animals, he despatched two young women, first Rosalie Osborne and later on Jill Donisthorpe, to the volcanoes. Neither of them, however, could claim to have met with much success in this hostile area.

When Dart heard about Leakey's initiative he became interested, and decided that he would go and see for himself what the possibilities were. Off he went to the little safari hotel, "Travellers Rest" in the small town of Kisoro, on the Uganda side of the Congo border.

It was the owner of Travellers' Rest, Walter Baumgartel, who had originally drawn Leakey's attention to the gorillas. Baumgartel was undoubtedly more interested in drawing visitors to his hotel than in scientific work – a fact he certainly would have vehemently denied – but his finances were at the time in a very rocky state and he was desperate. The relatively easy access to the gorillas conjured up in his imagination the dream of Kisoro as a world-famous centre for the study of such animals. Dart, who himself is an emotional imaginative man, was smitten with Baumgartel's enthusiasm and, soon after his return to Johannesburg, succeeded in persuading the University to grant him funds for the erection of a camp in the saddle between the two mighty volcanoes, Muhavura and Mgahinga, of a more permanent nature than the one used by Jill Donisthorpe.

It was now Dart's suggestion that I should apply for study leave and spend some time in Uganda to become acquainted with the gorillas. The leave was soon granted. I was given £500 to cover my expenses, and shortly before Christmas 1958 I set off. Unfortunately my attempts to study wild gorillas did not meet with much more success than my predecessors'. The unrest in the Congo, which later on resulted in the creation of the independent states of Zaire, Rwanda and Burundi, was accelerating, and my permits to wander unhindered across the border were cancelled. My freedom of movement was severely curtailed and I was virtually unable to work in Belgian territory.

The vegetation in the area in which I was permitted to work was dense, impenetrable bush, which made observations very difficult, if not impossible. Although I met the animals almost daily, I never saw them for more than a few moments at a time. Added to my frustration were the many tourists sent up by Baumgartel. They rarely had as much as the most elementary knowledge of wild animals. Some even complained that there were no paved footpaths through the jungle. The women usually came in colourful skirts and unsuitable high-heeled shoes. They were noisy and made work impossible.

Towards the end of my stay on the volcanoes, I had a violent attack of malaria, which resulted in a brief stay in a mission hospital in the Congo. In spite of all these difficulties, I did learn something which came in useful.

Shortly before I was due to break camp and leave Uganda, two distinguished zoologists, Schaller and Emlin, from Wisconsin, arrived. They, too, found that their permits to work in the border districts of the Congo had been cancelled, but with the help of some influential friends of Emlin's, they soon had them restored. I was glad, with my experience, to be able to help them to a good start.

While in Uganda, I was offered the post of Senior Lecturer in Zoology at Makerere. Because the political situation in South Africa made my position there precarious, I accepted. When, a few months later, I returned to Uganda, Emlin had left and Schaller was already busy on a very thorough study of the ecology of the mountain gorillas. This study was financed by the New York Zoological Society. The chairman of the Society, Fairfield-Osborne, asked me to keep an eye on Schaller and assist him when needed. During the first of the two years the study lasted, Schaller was helped by his wife Kay, who bravely endured much hardship. When the unrest along the border worsened and the safety of their lives was threatened, Kay moved over to us in Kampala, where she got a temporary job.

Schaller did a splendid job, and it was a joy to hear him relate his experiences. Once the gorillas were acquainted with him, they lost much of their fear, and it came to the point when young gorillas came and sat on the same low branch as he and watched him while he watched them. Most of Schaller's work took place in Albert National Park, close to the border towards Uganda. Unfortunately, I was not allowed to join him. Once, however, we went together to the Impenetrable (Kayonse) Forest in Uganda, where I had decided to build a hut on a mountain top. The hut was intended to serve as a base for research workers and especially for those working on the ecology and behaviour of gorillas. The money towards it was a present from the New York Zoological Society. Unfortunately, there were not many, besides Schaller, Professor Itani from Kyoto, Japan, and myself who made use of it.

Although my own study of the gorillas yielded very little of significance, the time I had with these animals was unforgettable, and my love for these impressive but peaceful animals was permanently aroused.

2.

A few years later, when I was sitting in my office in Nigeria, Bob Golding, my assistant, came in to tell me that some dangerous looking orientals were waiting outside with a pair of small gorillas which they offered for sale. To my great amazement Golding was right.

There were two small, miserable and frightened-looking gorillas, who showed all signs of having been maltreated. We estimated the male to be a little more than two years of age. He had large burns on his body. I asked the men how he had got such injuries. They said that he had tried to attack them, and to defend themselves they had hit him with a burning stick. The female was smaller and younger than the male – probably just over a year old – and was in a much better condition.

Gorilla

It was prohibited by law to trade in apes in Nigeria. I had several times been involved in attempts to stop the export of chimpanzees. Now I went straight to the phone to call the Minister for Natural Resources in Lagos, but as usual, the connection proved impossible. A call to the Danish game warden for Western Nigeria, Christian Brandt, was more successful. On many previous occasions he had given me valuable support. He had also played an important role in raising funds towards the upkeep of the small elephant Bodunrin. We agreed that I should try to detain the animals until I was able to get instructions from the Minister himself. The question of having the men detained by the police could wait. I had my instructions the following day: I should keep the animals until further notice. In case of difficulties or obstructions from the men, I should call the police and have them arrested. The police were informed by the Minister.

It was not surprising that the men felt disgruntled when I told them that I had been instructed to keep the animals until I received further orders from my superiors. I succeeded, however, in calming them down. The manner in which I handled the case may have seemed mean to many people, but there was nothing else I could do. It was always very difficult to communicate with Lagos, and, despite many requests, I did not have the necessary police authority.

During the next few months we were pestered by the traders who demanded their money for the animals, otherwise I would have to return them. The men became increasingly threatening, and Golding became afraid of the possible consequences. He began to suspect that the men might burgle the Zoo by night and steal the animals, or perhaps ambush him and insert a knife between his ribs. The latter he found a particularly worrying prospect!

We therefore got a police constable every night to guard the gorillas' cage.

The dramatic events visualised by Golding never materialised, but for several months I had one solicitor after another calling at my office, demanding that I paid for the gorillas. These gentlemen were always very polite when they arrived, but as soon as they discovered that persuasion was to no avail, they threatened me. They wanted me arrested for theft. When they began to threaten, I pulled out the letter from the Minister ordering me to have the men arrested if they caused any trouble. It always had a comic effect.

Golding's worries came to an end when finally the men were arrested and gaoled for an armed robbery.

We named the gorillas Aruna and Emadi. Golding, who was very keen to work with them, was put in charge as their foster-parent. It was obvious that they needed more care than I could give in the time at my disposal. Much time and patience was needed before the animals settled in to their new surroundings.

Golding was marvellous. During the first weeks he spent most of his time with his charges. They obviously needed someone who would love them. At one stage it looked as if they would not survive, and it took much care and patience before they recovered from the loss of their real mother. Luckily we had a young chimpanzee with whom they could play, and when Golding could spare the time, he took them out into the garden, where they could climb the trees, walk around and look at the other animals. Just like human children they liked to "visit the zoo". When they came to the terrapins they always climbed the low wall and sat down in the pool, splashing water over themselves.

A year after the arrival of the gorillas I left for America. While I was there, Golding succeeded in raising the money for a large new building for the big apes. Outside this were lovely, large playgrounds, which were surrounded by a moat with running water and filters, as in a swimming bath. At the opening of the new quarters, the keepers decided to have a swim. They jumped into the moat, and the gorillas, who could not swim, followed suit. Luckily they were rescued, but because they liked to play in the water where it was not too deep, it became necessary to put up an electric fence in the moat, so that they could not get out into the deep water. This fence caused them much amusement. They enjoyed pushing one another into it and short-circuiting the wires. Soon they also found out how to climb it without receiving shocks. Consequently we had to build the fence higher and step up the current.

I remember one day when Aruna, bigger now, had climbed over the fence and had a lovely time in the deep water side. When he saw me his face immediatley changed, and when I scolded him he waded to the end of the fence where it joined the wall. Here he proceeded, while supporting himself against the wall, carefully to climb back over the fence. He looked as crestfallen as a mischievous boy caught red-handed.

These two gorillas' fondness for water was unique. In the wild they are not seen to wade through a stream. They only cross where it is bridged by a tree trunk or some rocks.

Aruna and Emadi spent much of their time in the moat. They splashed and wrestled and ducked one another. They learned to swim a few strokes, copying their keepers. How long they could keep afloat without touching the bottom with their feet I never found out. Probably, their heavy build made it impossible for them to swim more than three or four strokes at a time.

It is likely that these two gorillas had learned from their keepers not to fear water. All monkey children learn from their mothers and from their superiors. It is possible that they, like human children, "felt big" when they found they could do the same as the adults. Aruna and Emadi did not have other gorillas, only Golding and their keepers to copy and to learn from. Given a scrubbing brush or a broom, they could scrub or sweep for hours, the same way they had seen their keepers do it. The chimpanzees and the drills could do the same, but no other monkeys did. It stands to reason that none of these animals had any idea why their cages were scrubbed, and their progress was haphazard.

The use of tools among apes and monkeys is only possible if the animals have had a chance, by experience, to learn what the tool can do. They must play with it and experiment with it before they can actually use it. Maco, the baboon, learned the use of sticks and stones in this way. She experimented when she had nothing else to occupy her – when she became bored. In the wild it is very rare for monkeys to make use of tools, but the apes often do. They may even invent their own simple tools.

Our gorillas in Nigeria liked to use stones as missiles. They seemed to enjoy the sight of people scattering in all directions when they hurled a rock at them. We removed all stones we could find from their yard, but then they started to dig for a new supply. When the soil proved too hard for their fingers, they used sticks for digging. One day we decided to dig over the whole area to remove what stones we could find to a depth of six to eight inches, after a keeper had been knocked out by one. But, even though we had removed all dangerous objects within easy reach, we did not curb the gorillas' desire to bombard the public. Having no more hard objects, they used whatever was available. Soon they discovered that nothing was better for creating panic among our visitors than a big lump of their droppings sent flying through the air.

3.

The closest relatives of the gorillas are the chimpanzees. The two apes have much in common in their behaviour, but I find that the chimpanzees lack the quiet dignity of the gorillas. They are the clowns and the jesters among the apes, and they love to draw the public's attention to themselves. I remember one young chimpanzee that was very unhappy and cried miserably when it could not perform in the arena and ride a bicycle together with the other chimps. Unfortunately, it had not yet learned which way to turn the pedals. I do not like to see chimpanzees dressed up and acting as clowns. It is undignified. But the chimps usually enjoy it and, to captives, it is undoubtedly better to act as clowns than it is to sit all day long in a soul-destroying cage.

Young chimpanzees can be extremely attractive and humanlike. Few people are able to resist them. When they grow up they become very dangerous. I had a young chimpanzee myself in my home for a short time. It was charming and amusing. It had its bed in a box on the verandah, and every morning when we woke up we could see it peeping through the window to see whether we were awake or still asleep. No sooner was the door open before he came in and up in the bed, where he snuggled down with us and began playing small games, similar to those engaged in by human children. It was this same chimpanzee we later used as a helper to get the gorillas through their initial depression. Later on he became very dangerous and a despotic ruler of his harem.

Chimpanzees are remarkably skilful at making and using simple tools in the same manner as the gorillas do it. Jane van Lawick-Goodall has described in detail how the wild chimpanzees fashion sticks so that they can be used for fishing termites out of their holes; how, by chewing leaves, they fashion wads with which they can sponge drinking water out of hollow trees; and how they use dry grass for cleaning their faces. Our gorilla female, Aruna, also cleaned her face with hands full of dry grass. Kortlandt has described and filmed wild chimpanzees clubbing a stuffed leopard with sticks.

The habit of chimpanzees in Zoos to bombard the public with whatever objects are at hand often causes concern. In Nigeria we had a chimpanzee named Janet, which was particularly keen on throwing things at people. To put an end to that, a heavy iron mesh was fastened over the bars of her cage. There was, however, a small hole left in the top, big enough for Janet to put a hand through. It always amused me to watch her from a hidden place close to her cage. She could be sitting on the floor with an expressionless, drowsy look on her face, waiting for a big crowd to gather outside. Then she would get up and slowly cross the floor. Unseen by the public, she would collect a good fistful of garbage and calmly climb to the top of the cage. There she would put her hand through the hole and suddenly, without warning, fling its contents on to the unsuspecting people. The victims would flee in all directions, while Janet gave vent to her joy with wild screams.

At the sound of the commotion, the keepers always rushed to the cage, shouting at the people, whom they suspected of having teased the chimpanzee. I am afraid that I was too mean to reveal to anybody what had happened, that it was not the onlookers who had teased Janet, but Janet who had teased the onlookers. After all, she too needed a little fun.

In one of the centres in America where they study the behaviour and the intelligence of chimpanzees, I walked one day alongside a cage housing several animals. Suddenly an object came flying through the air and, before I could get out of its way, it got stuck to my nose. It was one of the chimpanzees having fun decorating the most prominent part of my face with a lump of its own droppings. Although the distance was not very great, the aim had been remarkable. I tried to behave as if nothing had happened and continued my walk without changing pace. The lump was not removed until I was out of sight. This, however, did not fool the chimpanzees, who shrieked with joy.

For several decades, attempts have been made to teach chimpanzees to communicate by means of a sound language comparable to ours. It has always been my opinion that this is not possible, because the chimpanzees have neither the anatomical nor the physiological mechanism required for this.

To all monkeys, visual signs play a greater role in their communication than sounds. Sounds uttered at a short distance serve mainly to emphasise the meaning of their gestures, just as we use gestures to emphasise what we say. There are some exceptions. Warning and distress calls are always understood and it has been found recently that some vervets have different warning calls for different predators. A couple of decades ago, it was discovered that chimpanzees and gorillas were able to learn to communicate by using the American deaf-and-dumb language, in which there is no use of sounds.

Our ability to speak – to communicate with sounds without the use of signs – has developed at the expense of our ability to use our eyes to observe with. Once I made an experiment in which I used a young chimpanzee and a small group of children aged four to five years. In front of each subject I placed two identical bowls. Each of these bowls was covered with a lid, and on each lid was a simple drawing consisting of a few elements which were not the same in all drawings. If one of these elements was, for example, a circle, a reward was hidden under that lid. The chimpanzee and the children were to discover by trial-and-error which element gave the reward away. All subjects were shown without the use of words, what they were expected to do, and the chimpanzee and the children were treated in an identical manner.

It took less than ten minutes for the chimpanzee to discover what to look for. It worked at full speed and obtained a big reward. The children were slow, and it took a long time before they realised which figures were associated with the reward. One little girl never made it. If I had been a chimpanzee I might have drawn the conclusion that human children are less intelligent than young chimpanzees.

The cause of the children's inferior performance was, of course, not due to an inferior intelligence. It was a result of their acquisition of a highly developed mode of communication by means of sounds, which makes visual observation less important.

The big chimpanzees we had in the Zoo during my first stay in Nigeria were all animals who had been house-pets, and therefore thought of themselves as human beings. One of them was Penelope. She was a thoroughly humanised chimpanzee who, before she became a zoo-animal, was accustomed every night to going out with her owner to the club to have a Sundowner at the bar and a cigar. She loved smoking. Deliberately she would suck her cigar, inhale the smoke, and then slowly blow it out through her nose. She knew how to extract every possible pleasure out of her cigar. After her confinement in the Zoo, we often had a smoke together. I always gave her an unlit cheroot and then lit it for her while she sucked it. It was no good giving her the matches, because she found them so fascinating that she could forget all about the cigar. One by one she would carefully pick the matches out of the box and strike them until there were no more.

It happened, of course, that a chimpanzee sometimes got ill and needed medical attention. Our professor in veterinary science, Desmond Hill, had taught them that when they needed an injection they should come to the front of the cage and sit with their back against the bars. They would then get a prick in their shoulder and be rewarded with a lump of barley sugar. They liked that, and we had no trouble.

Once Penelope got very ill from what appeared to be food poisoning. Calling her was of no avail. She would not move from her sleeping mat. We discussed the problem, and it was finally decided that I should try to enter the cage and see what happened. Hill and the keepers would stand guard by the door, ready to come to my rescue should it be necessary. It is doubtful whether they could have done much in any case had things gone wrong. An adult chimpanzee is much stronger than a human and can be very dangerous. Luckily for me, I was well received by Penelope. She remained on her mat while I talked to her, stroked her and felt her forehead. When, finally, I had convinced myself that she would do no harm, I told Hill and the keepers to come in. She had her injection and a Coca-Cola with medicine in. She appeared to enjoy that. Hill ordained that for a week Penelope should have a daily drink with medicine, and I would have to give it to her because she seemed friendly disposed towards me. As it turned out, she gave me no trouble.

Penelope loved her daily Coca-Cola. When she needed no more medicine, she still demanded her drink. As soon as she heard me approaching her cage, she began screaming and hooting at the top of her voice, and, when I entered her door with the bottle in my hand, she hurried over to her mat, where she rolled up and moaned as if she was suffering from a severe belly ache.

Never before had I known that a Coca-Cola could be imbibed in so many ways. By using both lips she could make a funnel, through which I could pour the drink. Or she could make it from the lower lip alone. All the time she was experimenting. One moment she wanted me to empty the bottle into the middle of her mouth, the next through its corner. She always tried to make her drink last as long as possible.

It was remarkable that this animal, which we all regarded as extremely dangerous, became so friendly towards me. For several months I paid her my daily visit to give her a drink. It was not until I had to go away to Europe for a couple of months that my visits were brought to an end.

It is peculiar that some animals seem to understand when you want to help them. Once I visited a zoo where the Director was worried because their fine old male chimpanzee was ill. He asked me to have a look at the animal and give an opinion. There was no doubt that the animal was feeling very ill. It allowed me to examine it, although I was a stranger. I felt its forehead; it was hot. I looked at its throat; it was red and swollen. It was obvious that the chimpanzee had a very inflamed throat. I have since often wondered why this huge animal showed so much trust in me.

MAN AND ANIMALS

It is impossible for me to write about all my experiences with animals in these few pages. But I must point out that I have loved all the animals I have met, and would like to think that they have been fond of me. It always fills me with joy when an animal comes up to me to be patted or stroked, or when it in some ways indicates that it appreciates my presence. I am thrilled when an animal, after a long separation, recognises me and looks glad to see me. The reception I was given by several animals on my return to Nigeria, after having spent four years in America, can hardly be interpreted as anything else but a joyful recognition. What other reason could there be for the elephants to come to me and lay their trunks round my shoulders in an embrace? Could I explain the soft call of an old monkey, when she saw me at a distance, as anything else but the fact that she recognised me and wanted me to come to her? She did not behave like that towards other people. My daughter's annual meeting with the baboon Maco left no doubt that the animal recognised her and was glad to see her.

Many scientists are, in my opinion, much too afraid of expressing the view that animals may experience feelings and emotions akin to their own. It is a friendly act when a cat rubs itself against a person and purrs. By doing so, it leaves some of its own scent on the person, thus indicating that the person is acceptable. No-one can doubt that the dog (or the wolf) that jumps up and licks its owner, preferably on his face and around his ears, is glad to see him. Each animal species has its own ways of expressing its emotions. When it is done in a manner akin to our own we find it easy to understand. But it does happen that even closely related species express themselves by different means. Even among monkeys this may be the case. Not all monkeys express their affection by embracing and kissing, but when a monkey behaves like this towards us, we have no doubt about what it means.

Once I tried to teach a young adult male monkey to work a kind of slot-machine. It had to put a small stone in one hole in order to get a raisin from another one. Although I demonstrated several times what he had to do in order to receive his reward, for a long time he seemed to ignore the hole into which I dropped the stone. He only paid attention to the hole from which the raisin had appeared. He looked into it, poked his finger into it, and finally got into a fit of rage, shaking and tearing at the apparatus. When he realised that this was to no avail, he sat quiet for a brief moment before cautiously moving down to the floor. He picked up a stone, carried it up to the slot-machine, and dropped it into the correct hole. As soon as he had received the reward, he jumped over to me and flung his arms round my neck. I felt as if he was saying to me "Wasn't I a clever boy?" Of course I told him he was. His behaviour was so touching and so human. It reminded me of the behaviour shown by many human children after they have solved a difficult problem.

2.

When constantly surrounded by many animals, as I have often been, I cannot help noticing how much we humans have in common with them; much more than most people will admit. Friendship between humans and especially tame animals soon develops. But a friendship betweeen humans and wild animals may also develop.

The first time I really experienced and understood the feeling of comradeship with wild animals was in the Kalahari desert. There was something about the landscape, the gently rolling sand dunes, the soft pastel colours and the silence, which affected my emotions. It was teeming with animals, small as well as big ones. There were large herds of springbok, gemsbok and other antelopes feeding on the sparse vegetation. Sometimes a fox, a jackal or a hyena appeared close by. More rarely, a big cat. The days were hot, but the nights bitterly cold. Water was scarce and most of it undrinkable. Soon all the familiar worries were forgotten and only the present mattered. I became part of it all and I became an animal myself, sharing hardship with all those around me. We were equally exposed to hardship and danger. It did not matter that from time to time one of us should succumb. We were all partners in nature's great turn over, where death means the creation of new life. But, of course, I did not want to perish and be a hungry lion's meal. I just could not think about it and did not worry. Many times later on my travels in the wilds of Africa I had the same feelings.

3.

Many people have expressed surprise when hearing that I have always been unarmed. The truth is that I do not know how to handle a gun, and I believe I am safer without one. A greater expert than I once asked me if I felt no fear in the bush. When I answered that I didn't, he said 'Then you can be sure nothing will happen to you. You have no need of a gun.'

Of course there have been moments when I felt nervous, but I have never felt the same fear among animals as I have felt among my own kind. Thinking back, I cannot recall a single occasion on which I was in real danger, but when saying that I also have to admit that I have travelled less on foot than many other people. The most dangerous situation I encountered was possibly on a day when I, together with my trackers, followed some gorillas. In the dense vegetation we did not notice that we walked right into the middle of a herd of buffaloes. When one of my men suddenly startled a cow resting in the shade of a bush, the whole herd took fright and stampeded. There was only one small tree we could climb and it was immediately occupied by one of the trackers. The rest of us could do nothing but remain where we were, hoping for the best, while ton after ton of heavy buffalo came crashing past us, making the mountain reverberate like thunder. The old bull pinned us down to the spot for about twenty minutes.

4.

It may sound far-fetched when I say that animals may be able to sense one's friendly intentions, and yet I believe it to be possible. If I am right in this assumption, it should give the sympathetic person some protection when faced with a dangerous animal. I also believe that the animal's conflicting emotions are even more important to the outcome of such encounters. A strange animal will feel uncertain and may be torn between the urge to retreat and the one to attack. Prompted by its indecision, it will resort to a certain amount of bluffing. There is, of course, no reason to believe that the animal itself is aware that its behaviour serves to distract from its not always well concealed wish to withdraw. To the observer it is important to understand the underlying causes for this behaviour because it enables him to assess the situation and often predict the animal's reaction. But it is equally important that he is aware that the animal is also trying to sum him up and that it is very quick to notice any signs of tension, uncertainty or fear. He must be able to call the animal's bluff and outwit it. Actions opposite to what the animal might expect never fail to bewilder it. No movement must be uncertain and hesitant, and any outward sign of tension should be suppressed.

Once, a zoo keeper and I had to sex two young snow leopards. This could be done only by close examination. We did not want to endanger the life of such valuable animals by using any form of narcotics, so we decided to risk entering their cage without any such precaution. When we opened the door, both animals were crouching and hissing. A clear sign that they were frightened. As we approached they pressed themselves harder against the floor and their ears moved further and further back, until they were firmly pressed against the head. Our apparently calm approach terrified them. Having reached the animals, the keeper quietly took the leopards, one at a time by the scruff of the neck and lifted it for me to examine. This done, we left as calmly as we had entered.

On another occasion, the female jaguar escaped. As so often happens, the animal soon wanted to return to the safety of its cage, but it was unable to do so because the horrified keeper had locked the door, fearing that the male, too, might escape. It was not possible to get access to the door from the outside because the female blocked the narrow passage. The only possible way in was through the grotto. But that was where the male jaguar was. Finally, the head keeper, Nick, decided to jump down into the grotto. The male crouched and threatened a little, but was obviously afraid of Nick, who calmly crossed the grotto floor and got through the trap door to the inner cage. There he was then able to unlock and open the door to the passage where the female was.

Both situations could have been extremely dangerous, but because the animals were frightened and undetermined, it proved possible for Nick to bluff them enough to keep them at bay. Actions like this one take much self-control, and Nick was probably more frightened when he jumped into the grotto than the jaguar.

I remember a similar but far less dangerous experience in Africa. Once I met a huge male gorilla in a dense growth of bamboo. Only a minor clump of bamboo separated us. I think that we were both equally startled. The gorilla screamed at the top of his voice and bared his magnificent teeth. He started drumming the ground and hopping sideways. I wanted a photo of him dancing like that, and tried to push my way through the bamboo to a point where no vegetation could come between him and the lens of my camera. This, however, proved too much for him. He lost his nerve and fled before I could get my picture.

Gorillas are usually peaceful and timid, but they may attack when they feel cornered or believe that their females or young are threatened. Only on very rare occasions will they attack a man facing them. I have examined a number of men who bore scars from encounters with gorillas. They had all been accompanied by a dog and all of them had been bitten from behind in the arms, back, buttocks or the back of their thighs. No-one had scars in the front. It was obvious that all had been frightened and tried to flee. The only person I know of who has been attacked while facing a gorilla is the famous Rueben, my tracker. He was lucky to escape with only minor injuries.

5.

It has been a constant puzzle to me why so few people feel the urge to acquaint themselves with the wild and untamed nature. There are those who do not even care to watch a nature film. They prefer "Batman" and "Superman" or an exciting crime or war film in which the hero always escapes unharmed. Once I wanted to give a student of mine a happy experience and invited him to join me on a trip to the Kalahari. At noon, under a scorching hot sun we decided to have a rest under a thorn bush. While lying there in the sparse shade, I enjoyed the beauty surrounding us. The slender branches of the bush, with the shiny white thorns and the fine silvery-green foliage above us, was silhouetted against a clear blue sky. Next to me a single straw of grass, with purple top, rose out of the pale golden sand. The colours were delicate and gentle. I felt overwhelmed by their exquisite harmony and tried to draw the student's attention to it all. "You may keep your damned desert for yourself" was his reply. He was hot and thirsty and was longing to be back in the camp. Perhaps he was also hankering for town life back home. I felt terribly disappointed.

I was walking in London one day, after arriving back from a

holiday in Africa with my family. Everything looked grey and depressing. Nearly everybody seemed in a hurry and nobody looked happy and relaxed. A young couple passed me. They were less in a hurry than most people I saw, and they were holding hands. I could not help noticing them. He wore a leather jacket and trousers of somewhat exaggerated style, designed to show off his manliness. She wore a very short mini-skirt and shoes with very high heels. But most remarkable was her makeup. Her bright red mouth contrasted unpleasantly with her pale cheeks, each of which bore a large blotch of rouge. Her eyes were almost hidden beneath her long artificial eyelashes, bearing lumps of something resembling black shoe cream. Around them were deep shadows of mascara. I felt sorry for the couple. Their attempt at finding an identity and a little romance in a world so remote from the nature I loved was pitiful.

I mentioned this to a woman friend of mine. Her only answer was: "I prefer to look well-groomed to what you call natural."

OF WHAT USE?

Many years have passed since I walked in the garden with my grandmother. During these years I have seen many different but wonderful worlds. Perhaps I have succeeded in opening the doors for a few people here, though it seems that some are not really interested if it does not lead to material gains.

'What use are they?' asked a nice old lady many years ago when she saw me occupied with some insects.

'What use are you?' I burst out.

I hated the idea that animals should have been created for the sake of man. The mere thought was to me a sign of intolerable human arrogance. Luckily the old lady understood me when I explained what I meant, and took no offence.

My feelings are still the same today. I do not like the attitude which has prevailed since the days of Francis Bacon. For him nature was there to serve mankind. The goal of the scientist was that of making her a "slave". She should be "put into constraint" and her secrets "tortured out of her." I am frightened of the inevitable outcome if the world continues to ignore the warnings given by ecologists. My old revulsion against the use of poisons as a means of pest control is still with me, for poisons are rarely specific in their effect. They do not distinguish between useful and harmful animals. And what do we know about the sufferings they inflict? Are we right in assuming that a worm, a snail or an insect is incapable of suffering as much as we are because it possesses a simpler nervous system than ours. I would like to think so, but what do I know?

There can be little doubt that a higher animal such as a rat dying after eating bait poisoned with strychnine or thallium suffers a horrible death, but people are usually quite insensitive to the fact. If, by accident, their pet dog or cat should happen to eat the poison and go through the same pains, they would get very upset. This peculiar difference in attitude to suffering was highlighted when, on a hunt, the hounds mistook a cat for the fox and tore it to pieces. During the following days the newspapers were full of letters from readers, lamenting the cruel death of the cat and blaming the organisers of the hunt. Few poeple would have felt sorry if the sufferer had been the fox.

Man has an urge to know and to understand which is foreign to other animals. The old curse from Adam's fall is always with him. He searches for knowledge and asks: "Why, how, and of what use?" The greatest question to him is life itself and its meaning. It was really to this last question the old lady wanted an answer when she asked about the purpose of my insects. It is a question to which there is no answer, because the meaning of life is life itself.

What I was actually doing when the old lady asked the question was: I was trying to find out how some solitary bees found their way into the flowers and how they collected pollen. My observations and experiments were perhaps not in themselves of very great importance, but they were some of the many small building stones which, together with countless other studies, have contributed to our present understanding of the pollination of flowers. Not all insects are equally good pollinators and some seem to go at random from one kind of flower to another. Others visit only one kind of flower and stay with that, as long as it yields well, and thereafter, when it declines, shift to another kind. There are also insects which have specialised in visiting just one or only a few different flowers. We may even find that the flower and the insect are both adapted to one another so that only one species of insect can pollinate one species of flower. Knowledge of this kind is useful to plant breeders who are dependent on insects. It enables them to take precautions which ensure a higher yield.

Unfortunately, the use of insecticides has sometimes had a catastrophic effect on the insect population. It is no good killing the pests, when at the same time we kill their enemies, and, on top of that, all the insects which should pollinate the flowers.

Nearly all my life I have had the company of animals. This has given me endless joy, a joy I wish other people could experience. Of all animals, dogs and cats are probably those that make the best and most rewarding pets, but there is quite a range of other suitable animals. I have had pets belonging to a great variety of species. Most of them have been brought to our house as young orphans, and I have taken charge of them in the hope of saving their lives. I do not approve of trapping and capturing wild animals for the sake of making pets of them, but I do understand the thrill they may give their owner. Most of them need expert handling. Many are unreliable and even dangerous and, more often than not, they become a disappointment to their owner. Good pets can enrich their owner's life, and they make good company for the lonely and elderly.

Most people who have known and enjoyed the unrestrained wild animals in their natural habitat tend to regard zoos as abominations. I cannot claim to be very fond of zoos myself, but I regard them as valuable educational tools. It was for this reason I insisted on developing the University of Ibadan Zoo. Many people who cannot travel enjoy seeing the animals in the zoos, and there they gain a first-hand experience, which, when supplemented with the many excellent nature films, helps them to understand the animals as they are in their natural surroundings. Much joy may result from such an understanding.

When I made the decision to develop the Ibadan University Zoo, my aim was a two fold one. I wanted to give a very large urban population the joy of seeing wild animals they would otherwise have little or no chance to see. Moreover, I wanted to create an interest in the preservation of the country's quickly vanishing wild life. The Zoos all over the world have adopted similar ideas.

Besides having developed into educational centres, most bigger zoos are also centres where much good and valuable research takes place. Animal husbandry and veterinary science have benefitted from the experiences gained from a greater variety of animals than conventional livestock could provide.

Many people may not approve of my sentiments for nature as outlined here, and of my reverence for life. My approach to life may appear unrealistic to others. To me it seems beyond doubt that mankind would benefit from a closer contact with and a better understanding of nature.

We tend to regard ourselves as being apart from nature, when we are actually created to be a part of it. This idea, I fear, may in the future be the greatest obstacle to our survival. We must learn that we cannot take without giving. Co-operation must replace domination.